future arquitecturas s.l. 编

未来建筑竞标 中国 第1辑

空间隙地

FUTURE ARQUITECTURAS COMPETITIONS CHINA I
THE SPACES IN BETWEEN

图书在版编目（CIP）数据

未来建筑竞标. 中国. 第1辑：空间隙地：汉英对照 / 未来建筑编. 一杭州：浙江大学出版社，2010.11
ISBN 978-7-308-08048-4

Ⅰ. ①未… Ⅱ. ①未… Ⅲ. ①建筑设计—世界—现代—图集 Ⅳ. ①TU206

中国版本图书馆CIP数据核字（2010）第204124号

浙江省版权局著作权合同登记图字：11-2010-118号

未来建筑竞标 中国 第1辑 空间隙地

FUTURE ARQUITECTURAS COMPETITIONS CHINA I THE SPACES IN BETWEEN

future arquitecturas s.l. 编

责任编辑：黄娟琴
文字编辑：李峰伟 张凌静
封面设计：future arquitecturas s.l. 未来建筑
出版发行：浙江大学出版社
（杭州天目山路148号 邮政编码：310007）
（网址：http://www.zjupress.com）
印　刷：杭州多丽彩印有限公司
印　张：15
开　本：889mm×1194mm 1/8
字　数：192千
版印次：2010年11月第1版 2010年11月第1次印刷
书　号：ISBN 978-7-308-08048-4
定　价：120.00元

浙江大学出版社发行部邮购电话：88925591

美好心愿
当代空间结构

建筑师Jose Antonio Corrales于2010年7 25日逝世。生前在其马德里圣费尔南多皇家艺术学院院士的就职演说中，他指出了现代性的关键之处，并引用了Le Corbusier的话作为开场白："现代性的关键在于注视、观察、领会、想象、发明与创造"。

他结合直觉、道德规范和富有远见的想象发表了对建筑艺术关系的演说，使得其通过项目、竞争和构思来寻求当代空间结构，这就是他在艺术上所作的点睛之笔。他的技艺之基在于对图像世界的否定，相反却领悟了其真正的意义。

"建筑首先是供使用、居住和享受的，而非仅供研究的符号。"
Jose Antonio Corrales

他对现代性的观点与生活在伦敦的西班牙艺术家Angela de la Cruz如出一辙，其雕刻和绘画技艺使其艺术变得概念化，且富有当代气息，而他的观念同样又与德国哲学家Theodor Adorno相同，对当代艺术的观点是"工业化促成很多东西走向成熟，而艺术则将其照单全收。"

诗歌，作为哲学和尊严的源头，突出了他极富诗意的艺术技巧的精华。其中，他从一个相反的视角来看待美与善，正如诺贝尔文学奖得主Gabriela Mistral所言，这也是合乎伦理道德的。Gabriela Mistral于1957年在纽约逝世，但他对于人们的艺术教导却从未停止过（智利于2010年庆祝其独立两百周年）。

"建筑让事物变得永恒，变得伟大，令人崇敬。"
Ludwig Wittgenstein

GOOD INTENTIONS
The contemporary construction of space

In his inaugural speech as Academician of the Royal Academy of Fine Arts of San Fernando in Madrid, the architect Jose Antonio Corrales, who died on July 25, 2010, indicated the keys to modernity, using words of Le Corbusier as a prologue: "The key is to watch, observe, see, imagine, invent, create".

His defence of the artistic relationship of architecture, together with his intuition, ethics and ambitious imagination led him to search for the contemporary construction of space through his projects, contests and ideas, which have been key to his art. His technique is based on the denial of the world of images, which acquired its real meaning fo· its opposite.

"Architecture is first an object to use, live and enjoy than a symbol to observe"
Jose Antonio Corrales

His concept of modernity is in tune with the Spanish artist based in London, Angela de la Cruz, whose sculptural implementation mixed with his painting technique turns her art into conceptual and contemporary and, in tune with the German philosopher Theodor Adorno, whose concept of modern art is "that which absorbs everything that industrialization in the dominant forms of production has managed to transform into mature".

The poetry as a source of philosophy and dignity marks the essence of his poetical technique, considering the beauty and the good into a reverse order as they are ethical as expressed by the Literature Nobel Prize Gabriela Mistral, who died in New York in 1957 but who never ceased to be artistically didactic. (Chile celebrates the Bicentennial of the independence in 2010)

"Architecture makes something endless and sublime"
Ludwig Wittgenstein

浙江大学建筑工程学院现有规划、建筑、土木、水利四大学科，基本涵盖了国家基本建设领域的全部学科，涉及到建筑、市政、交通、水利、铁道、港口与海洋工程等主要产业领域。在学科上具有良好的互补性和交叉性，为学科发展奠定了良好的基础。

浙江大学建筑设计研究院始建于1953年，是国家重点高校中最早成立的甲级设计研究院之一。业务范围有高层、超高层的大型办公、宾馆、商业综合体、行政办公楼；学校校园规划与设计；影剧院、图书馆、博物馆等文化建筑；居住区规划与设计；体育建筑；医院类建筑；城市设计；智能建筑设计、室内设计；风景园林与景观设计；市政公用工程；岩土工程；幕墙设计；古建筑和近现代建筑的维修保护、文物保护规划等。

The College of Civil Engineering and Architecture of Zhejiang University has a wide coverage of the fields of national capital construction including: building design and construction, municipal engineering, transportation, water conservancy, railway and harbor and offshore engineering. And it forms a construction with Regional and Urban Planning, Architecture, Civil Engineering and Hydraulic Engineering to have the integration of production, learning and research.

Architectural Design and Research Institute of Zhejiang University was founded more than half a century ago in 1953, it has been one of the earliest Grade-A design institutes established among state key universities. The business covers high-rise or super high-rise large office, hotel, business complex and administrative office building; planning and design of campus; cinema, library, museum and other cultural buildings; residential community planning and design; construction of sports facilities; construction of hospital; urban design; intelligence architectural design, indoor design; landscape architecture and landscape design; municipal public engineering; geotechnical engineering; curtain walling design; maintenance and protection of ancient building and modern building, planning of cultural relics protection, etc.

宁波帮博物馆 · 宁波
Ningbo Bang Museum

董丹申 Dong Danshen · 胡洋 Hu Yang · 秦敏 Qin Min

三清山博物馆 · 江西
San Qing Shan Museum

沈济黄 Shen Jihuang · 方华 Fang Hua · 叶长青 Ye Changqing

浙江大学紫金港校区西区规划 · 杭州
Program for West Area of Zi Jing Gang Campus Zhejiang University

董丹申 Dong Danshen · 杨易栋 Yang Yidong · 徐若峰 Xu Ruofeng · 黄廷东 Huang Tingdong

三堡排涝工程建筑及景观方案 · 杭州
Drainage Project Construction and Landscape Plan of Sanbao

王竹 Wang Zhu · 唐君 Tang Jun · 陈宗炎 Chen Zongyan · 郭文东 Guo Wendong

西湖博物馆 · 杭州
West Lake Museum

余健 Yu Jian · 曾勤 Zeng Qin · 黎冰 Li Bing · 王雷 Wang Lei · 沈济黄 Shen Jihuang · 沈金 Shen Jin · 裘涛 Qiu Tao · 刘晓梅 Liu Xiaomei · 卢德海 Lu Dehai · 曹志刚 Cao Zhigang

宁波帮博物馆 · 宁波

Ningbo Bang Museum · China

董丹申 Dong Danshen · 胡洋 Hu Yang · 秦敏 Qin Min

竞标方案 competition proposal

设计首先从能够充分体现宁波帮共性的地方寻找切入点，而这一共性，究其根本就是“宁波”这一共同故乡。在这里，设计找到了能够承载宁波帮传奇与精神的舞台——岸。它代表着土地和家园，象征着丰饶美丽的故乡，象征着大海所孕育出的独特商帮文化。博物馆之美，一是在于自身，一是在于其内部展品的珍贵，两者相得益彰。建筑朴实有力的体形集中体现了理性、唯实、简约的建造精神，外实内虚的设计使建筑体现了强烈的形体张力，简约自由的形态使博物馆成为与环境浑然一体的建筑景观。层叠的室外平台以及大尺度的台阶，使建筑成为一组具有高度可达性的连续体量，创造丰富的建筑人文景观。

The museum is designed based on the general character of the people of Ningbo who are connected by their common hometown of "Ningbo". Here, the design of Bank, as a platform exhibiting the legend and spirit of the people of Ningbo, symbolizes the land and homestead, a rich and beautiful hometown, and a unique commercial culture with oceanic characteristics. The beauty of a museum lies behind both the entity and the valuable exhibitions inside. The simple and compacted structure reflects the spirit of sense, realism and conciseness; the design, featuring external pragmatism and internal style, intensively manifests a stretchy force; the concise and free structure makes the museum stand out in harmony with the environment. The outdoor terraces and the large steps contribute to an accessible architectural cascade, which acts as a profound architectural and humanistic view-point.

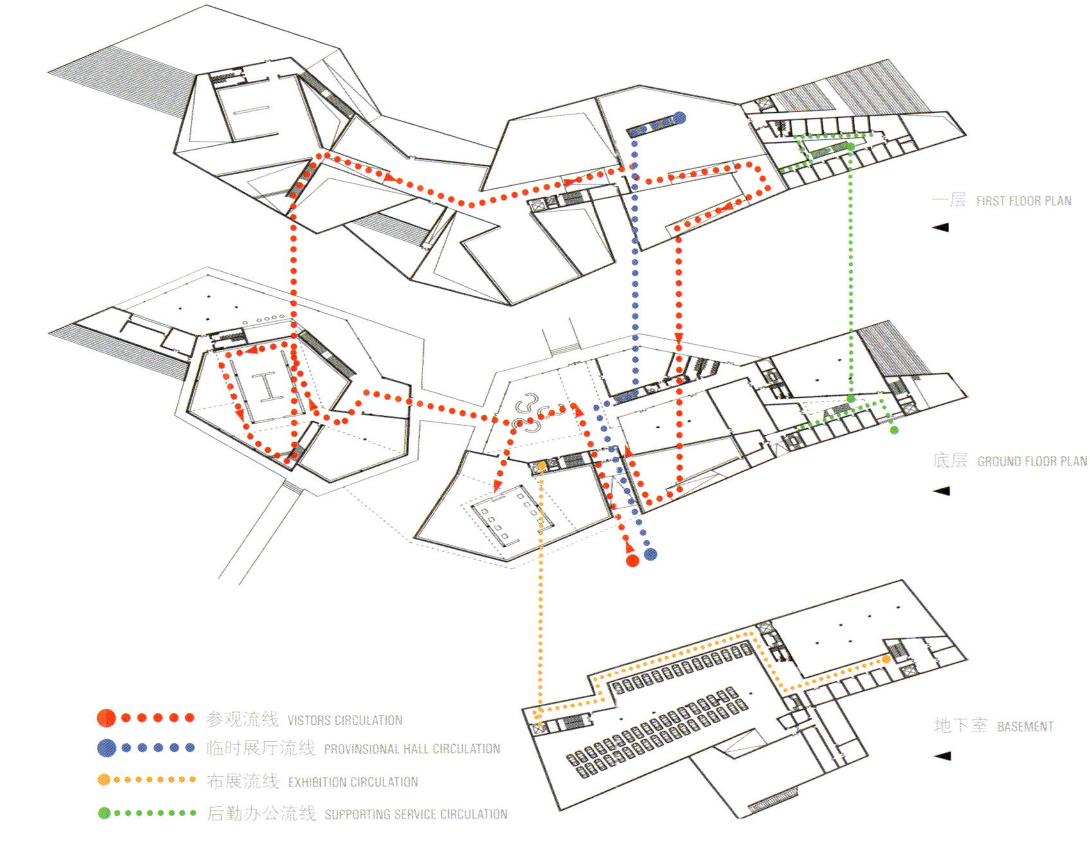

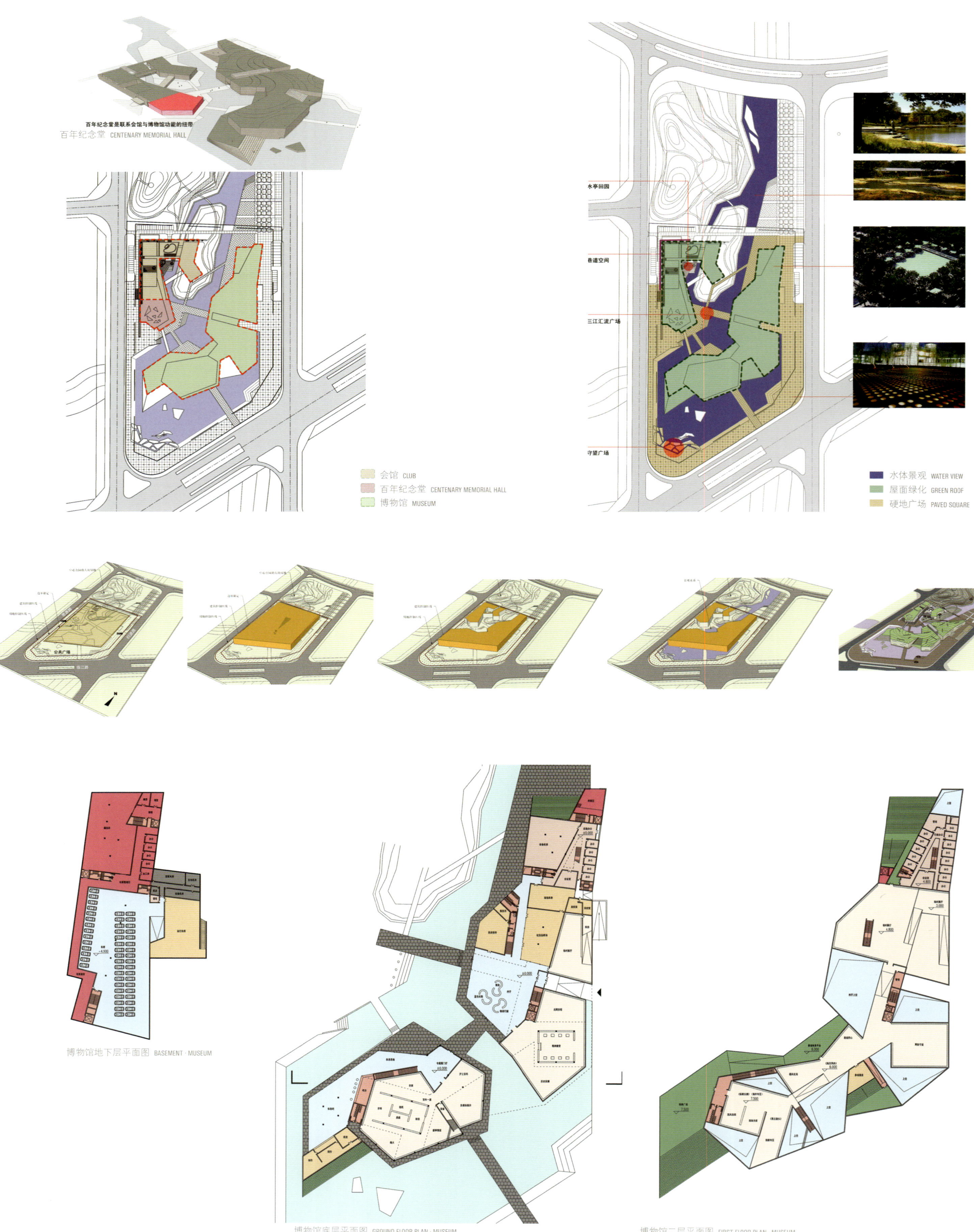

博物馆地下层平面图 BASEMENT · MUSEUM

博物馆底层平面图 GROUND FLOOR PLAN · MUSEUM

博物馆二层平面图 FIRST FLOOR PLAN · MUSEUM

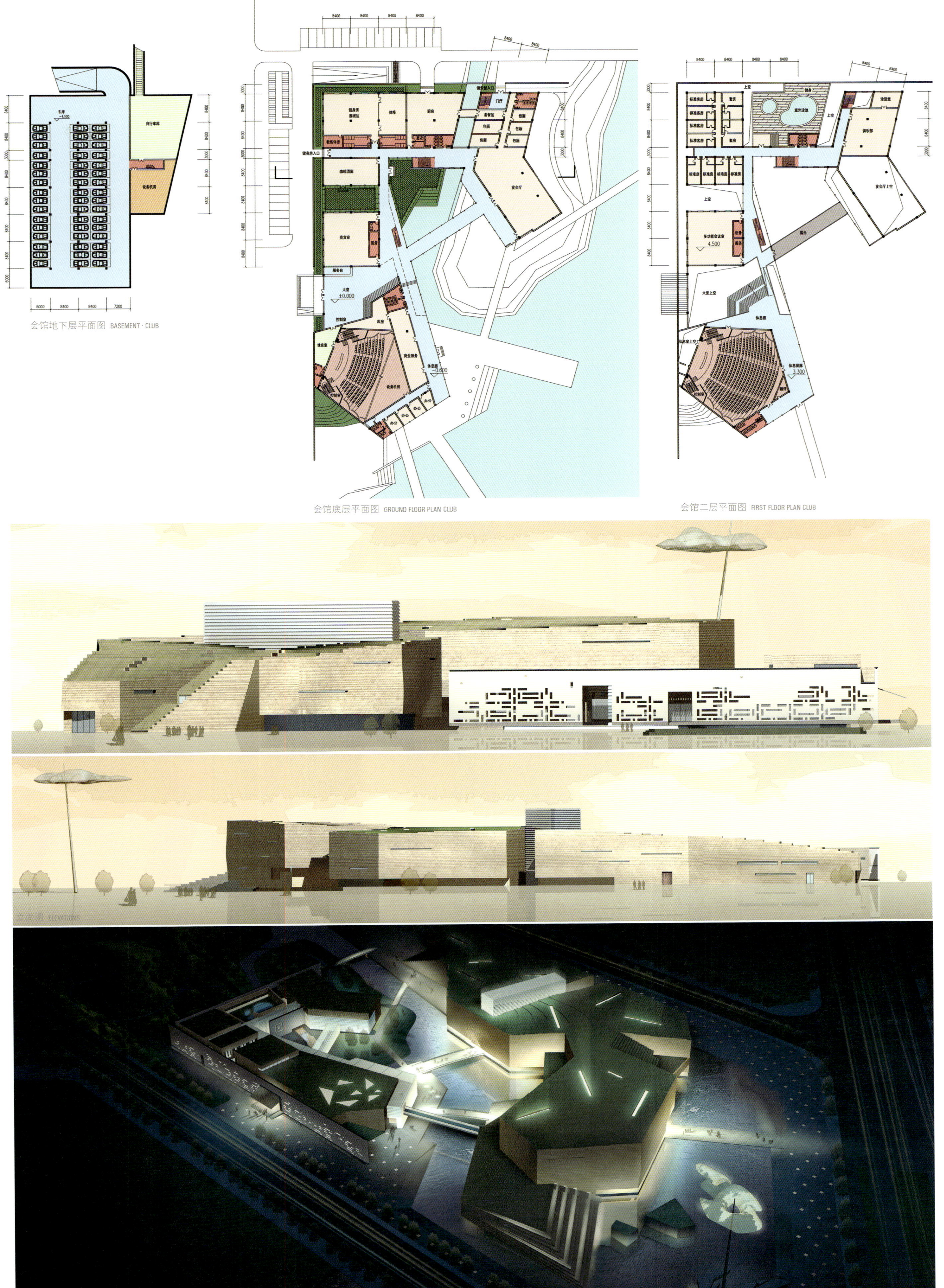

会馆地下层平面图 BASEMENT · CLUB

会馆底层平面图 GROUND FLOOR PLAN CLUB

会馆二层平面图 FIRST FLOOR PLAN CLUB

立面图 ELEVATIONS

三清山博物馆 · 江西

San Qing Shan Museum · China

沈济黄 Shen Jihuang · 方华 Fang Hua · 叶长青 Ye Changqing

竞标中标 competition winner

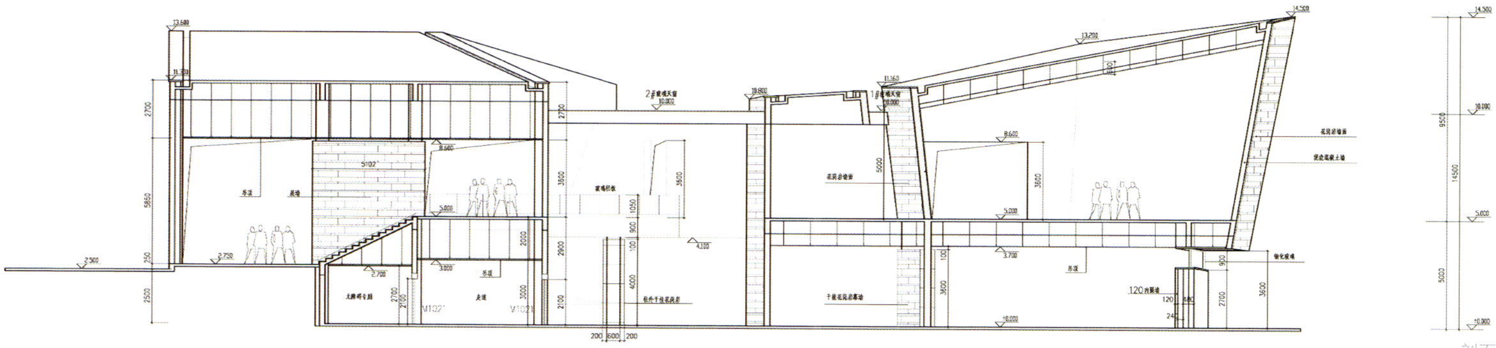

剖面图 A-A SECTION A-A

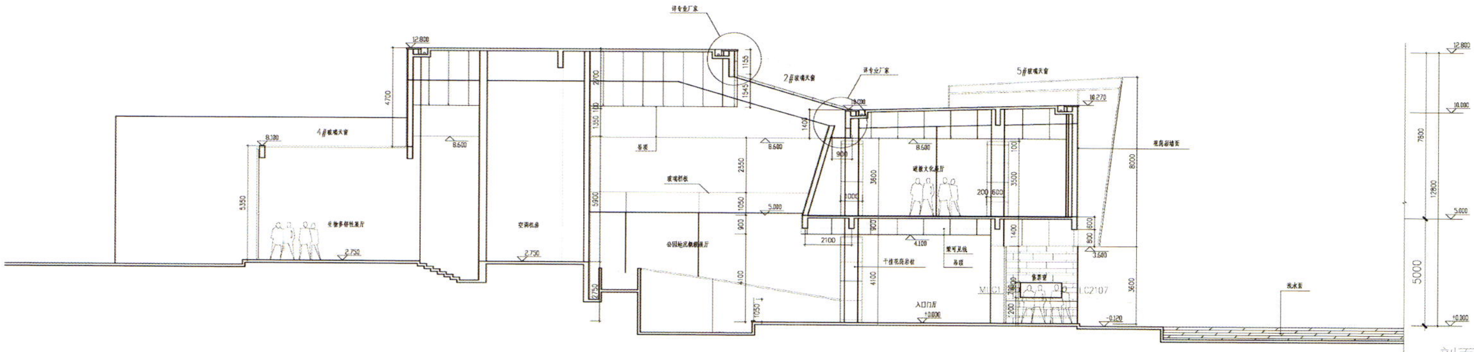

剖面图 B-B SECTION B-B

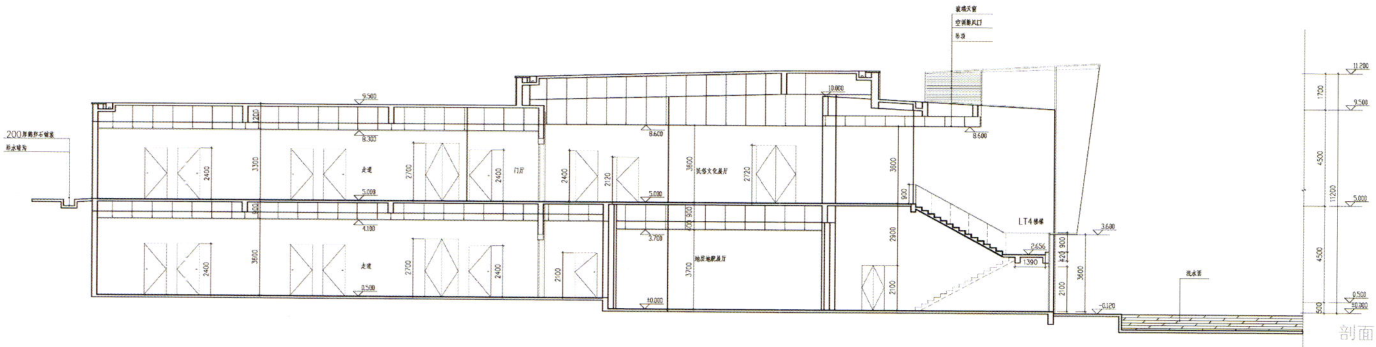

剖面图 C-C SECTION C-C

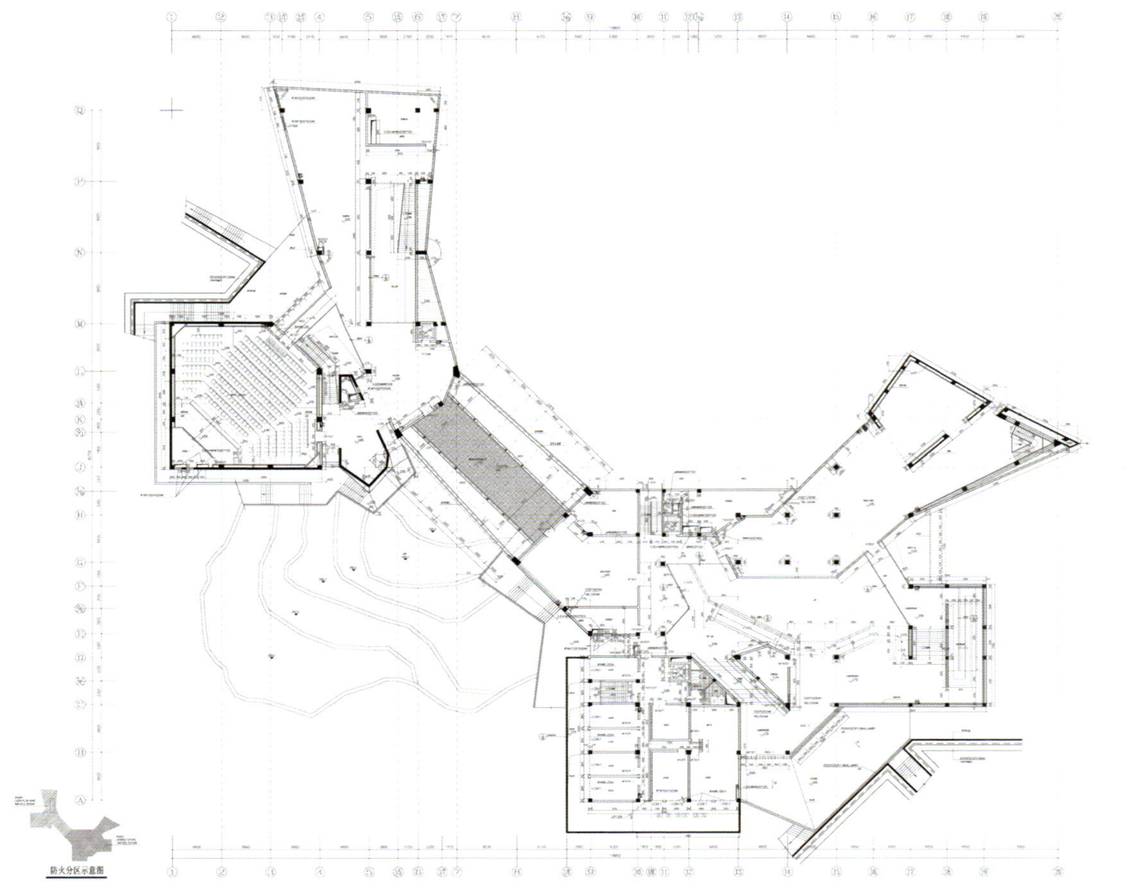

平面图 +5.00 FLOOR PLAN +5.00

三清山博物馆设计构思试图抓住三清山地质地貌形成中的海浸、板块之间的挤压、拉伸、走滑、切拉、断裂运动的某个瞬间，将其固化为展示各种地质运动力量的建筑片段，由此拼合成的博物馆形体犹如天工造化，充满张力并显示其独特内涵。

Sanqingshan Museum tries to base its design on some moments of marine movements, plate jams, plate stretching, tension and fracture of Sanqingshan Mountain, in an attempt to exhibit various geological movements. The museum, made up of d fferent architectural sections, is of superb exquisiteness, great tension and distinctive interior charm.

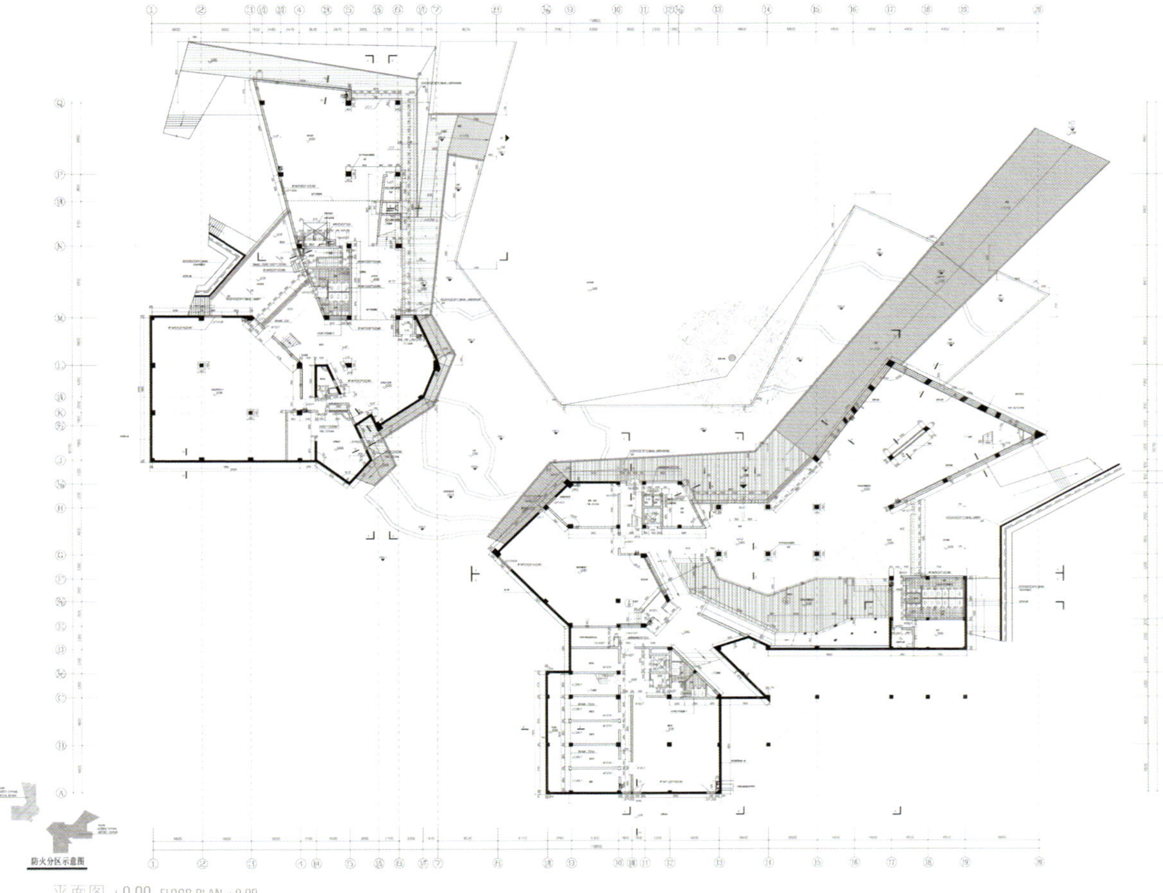

平面图 +0.00 FLOOR PLAN +0.00

浙江大学紫金港校区西区规划·杭州

Program for West Area of Zi Jing Gang Campus Zhe Jiang University· China

董丹申 Dong Danshen · 杨易栋 Yang Yidong · 徐若峰 Xu Ruofeng · 黄廷东 Huang Tingdong

竞标中标 competition winner

图书馆 LIBRARY

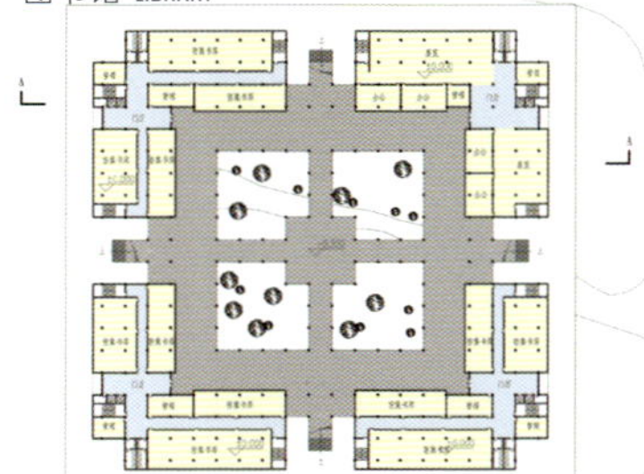

一层平面图 GROUND FLOOR PLAN

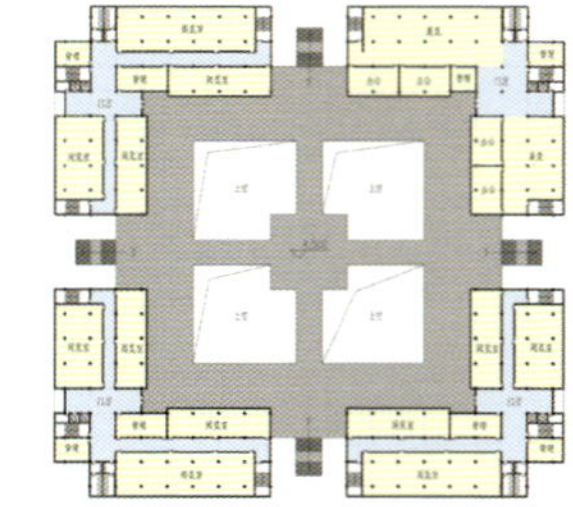

二层平面图 FIRST FLOOR PLAN

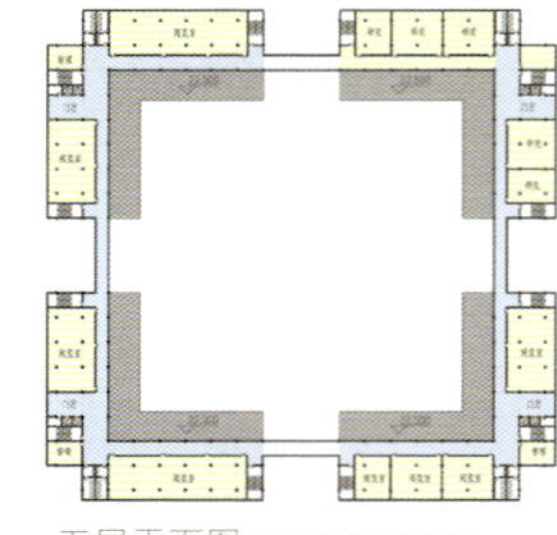

五层平面图 FOURTH FLOOR PLAN

总平面图 GENERAL PLAN

文科类学院 FACULTIES OF HUMANITIES

标准层平面图 TYPICAL FLOOR PLAN

"多心复环"的规划思路，解决超大型校园的功能要求和交通问题，营造尺度宜人的校园空间。对"大学园林"的内涵作拓展延伸，充分保护和利用基地内的原生湿地，体现江南水乡的环境特色。完成由东区向西区发展的自然过渡，构筑东西区一体化的生态校园环境。针对西区以研究型教学为基本运作的功能特点，探索适应学科组群发展的组团型规划及建筑模式。

The planning guideline of the "Multi-center, multi-ring" can solve the functional and traffic needs of the super-large campus, creating pleasant campus environment. The extension of the "campus garden" is intended to protect the primitive wetland of the site, reflecting the environment features of the south of the lower reaches of The Yangtze River. The design tries to complete the natural transition from the east district to the west district, building up an ecological campus environment which integrates both districts. The design explores the cluster-type planning and the architectural rules for groupped development.

一期 STAGE 1

二期 STAGE 2

远期 FUTURE

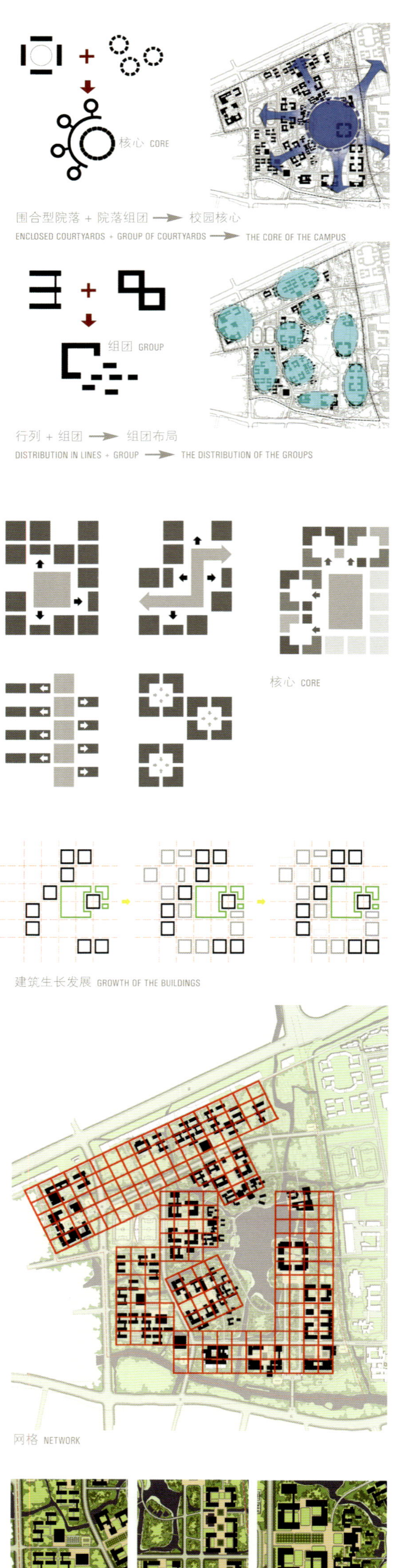

1-1 主校门 MAIN GATE
1-2 入口主轴线 MAIN AXIS FROM ENTRANCE
1-3 中心景观湖面 CENTRAL LAKE VIEW
1-4 试验农田 FARM LAB
1-5 试验农田2 FARM LAB 2
1-6 体育活动场地 SPORTS PLAYGROUND
1-7 预留发展用地 SITE FOR FUTURE EXTENSION
1-8 预留轻轨地下站点 SITE FOR FUTURE SUBWAY STATION
2-1 文学院楼群 BUILDINGS OF FACULTIES OF HUMANITIES
2-2 理学院楼群 BUILDINGS OF SCIENCE FACULTIES
2-3 工学院楼群 BUILDINGS OF ENGINEERING FACULTIES
3-1 研究平台交叉中心 RESEARCH PLATFORM CENTRE
3-2 公共教学楼 CLASSROOM BUILDINGS
3-3 行政中心 ADMINISTRATION CENTRE
3-4 艺术博物馆 ARTS MUSEUM
3-5 校史档案博物馆 MUSEUM OF HISTORY ARCHIVES
3-6 图书馆 LIBRARY
3-7 研究生教育中心 CENTRE OF MASTERS' EDUCATION
3-8 留学生中心 STUDENT ABROAD RESOURCE CENTRE
3-9 对外交流中心 CENTRE FOR FOREIGN COMMUNICATION PROGRAM
3-10 大型动物实验用房 FAUNAL LAB

4-1 博士后/专家公寓 APARTMENTS FOR POSTDOCTORS AND EXPERTS
4-2 学生公寓 APARTMENTS FOR STUDENTS
4-3 学生食堂 STUDENTS' CANTEEN
4-4 学生服务中心 SERVICE CENTRE FOR STUDENTS
4-5 后勤附属用房 SUPPORTING SERVICE FACILITIES
4-6 试验农田用房 FARM LAB BUILDING
4-7 会堂 AUDITORIUM
4-8 体育活动用房 SPORTS AND EVENTS
5-2 医学中心 MEDICAL SCIENCE CENTRE
5-2 医学中心发展用地 SITE FOR FUTURE EXTENSION OF MEDICAL SCIENCE CENTRE

东区 EAST

西区 WEST

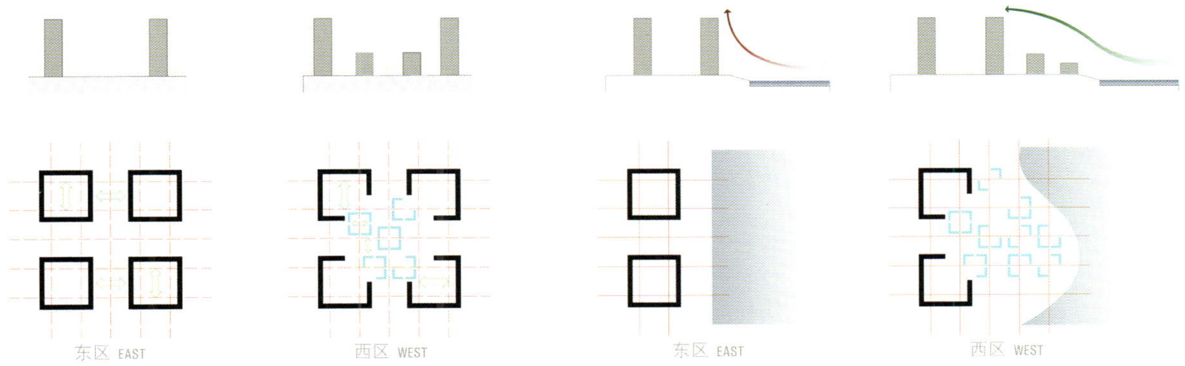

建筑之间 BETWEEN BUILDINGS

东区 EAST

西区 WEST

建筑与水 BETWEEN BUILDINGS AND WATER

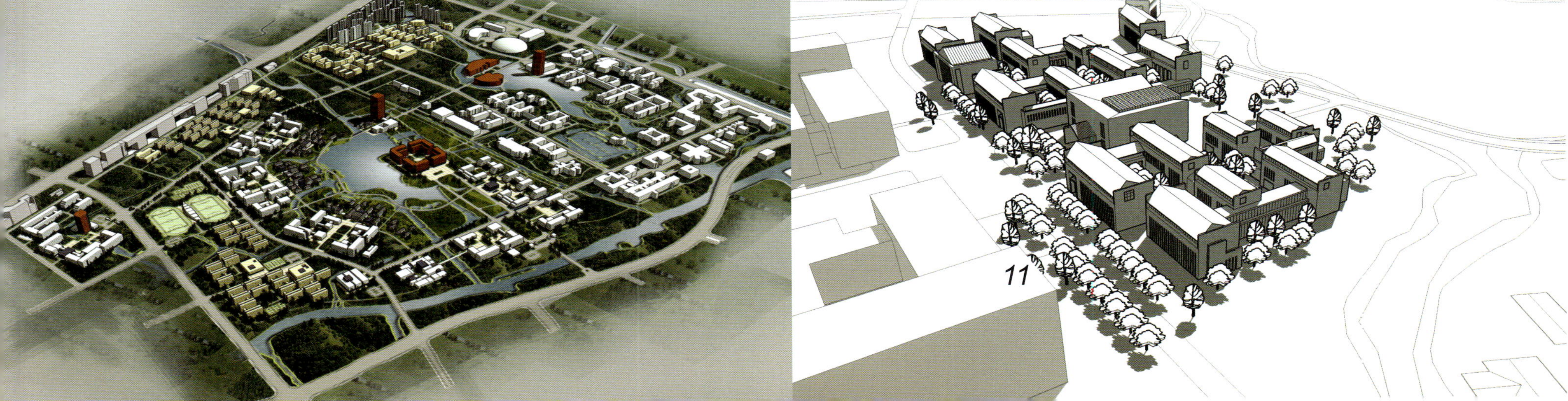

三堡排涝工程建筑及景观方案·杭州

Drainage Project Construction and Landscape Plan of Sanbao · China

王竹 Wang Zhu · 唐君 Tang Jun · 陈宗炎 Chen Zongyan · 郭文东 Guo Wendong

国际竞标方案 international competition proposal

本项目旨在利用现有的三堡排涝泵站上部厂房和管理用房，综合考虑接待、观光、展览等需要，在用地范围内辅以独特的景观设施，使其成为杭州水力标志性建筑及地块特色建筑。它将是一个给予人方向，驶向未来的帆船，也是在设计方案中强化利用的地块特征优势。设计采用空间箱形钢结构体系，使其既可以有大的出挑，又能够达到结构的自平衡，同时还采用双层LowE玻璃，外层铺设太阳能集热板，达到通风、隔热、保温与景观一体化的效果。

This project aims to build a hydraulic landmark building in Hangzhou with the block features by using the existing plant and the management space of Sanbao drainage pump station, with a unique landscape facility functions such as reception, sightseeing, exhibition, etc. It is to be designed as a future-oriented sailing boat that leads the people to a determined direction, and also concentrates on the usable advantages of the block features. It adopts a spatial box steel structure system, with not only an outstanding appearance, but also a self-balance structure. Two-layer Low-E glass and solar panels are used on the exterior layer, promoting an integration of ventilation, heat insulation, heat preservation and landscaping.

总平面图 OVERALL PLAN

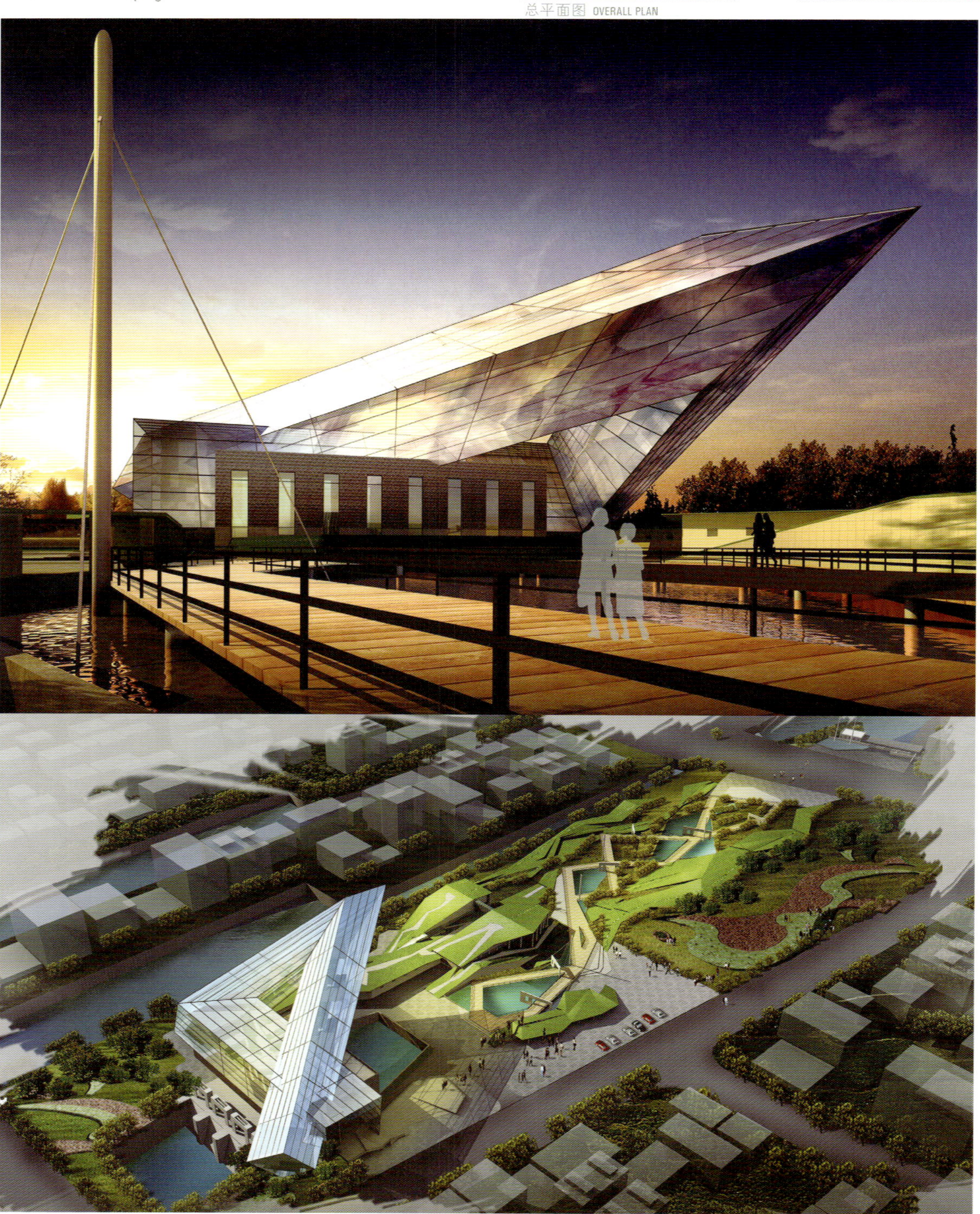

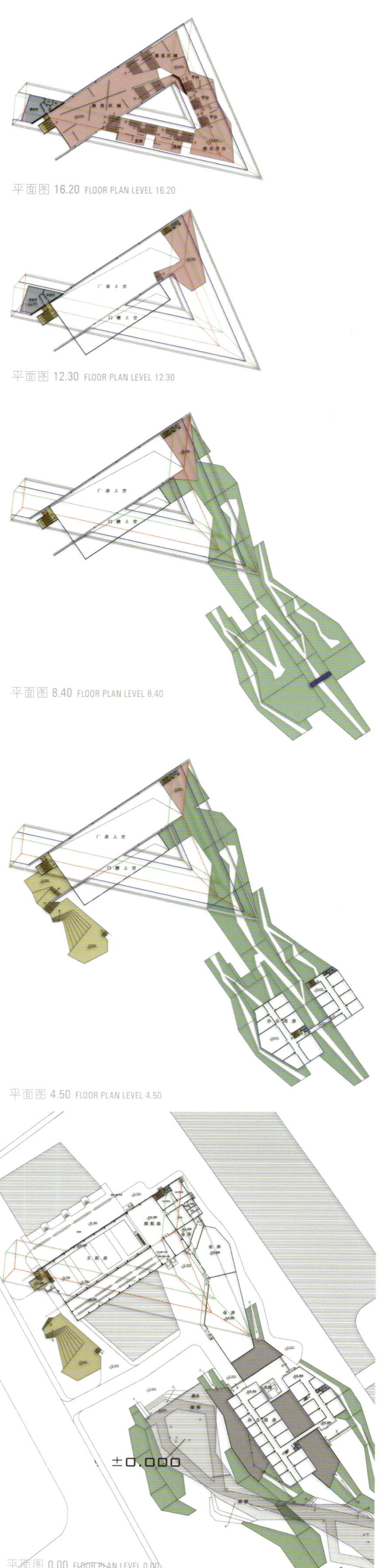

平面图 16.20 FLOOR PLAN LEVEL 16.20

平面图 12.30 FLOOR PLAN LEVEL 12.30

平面图 8.40 FLOOR PLAN LEVEL 8.40

平面图 4.50 FLOOR PLAN LEVEL 4.50

平面图 0.00 FLOOR PLAN LEVEL 0.00

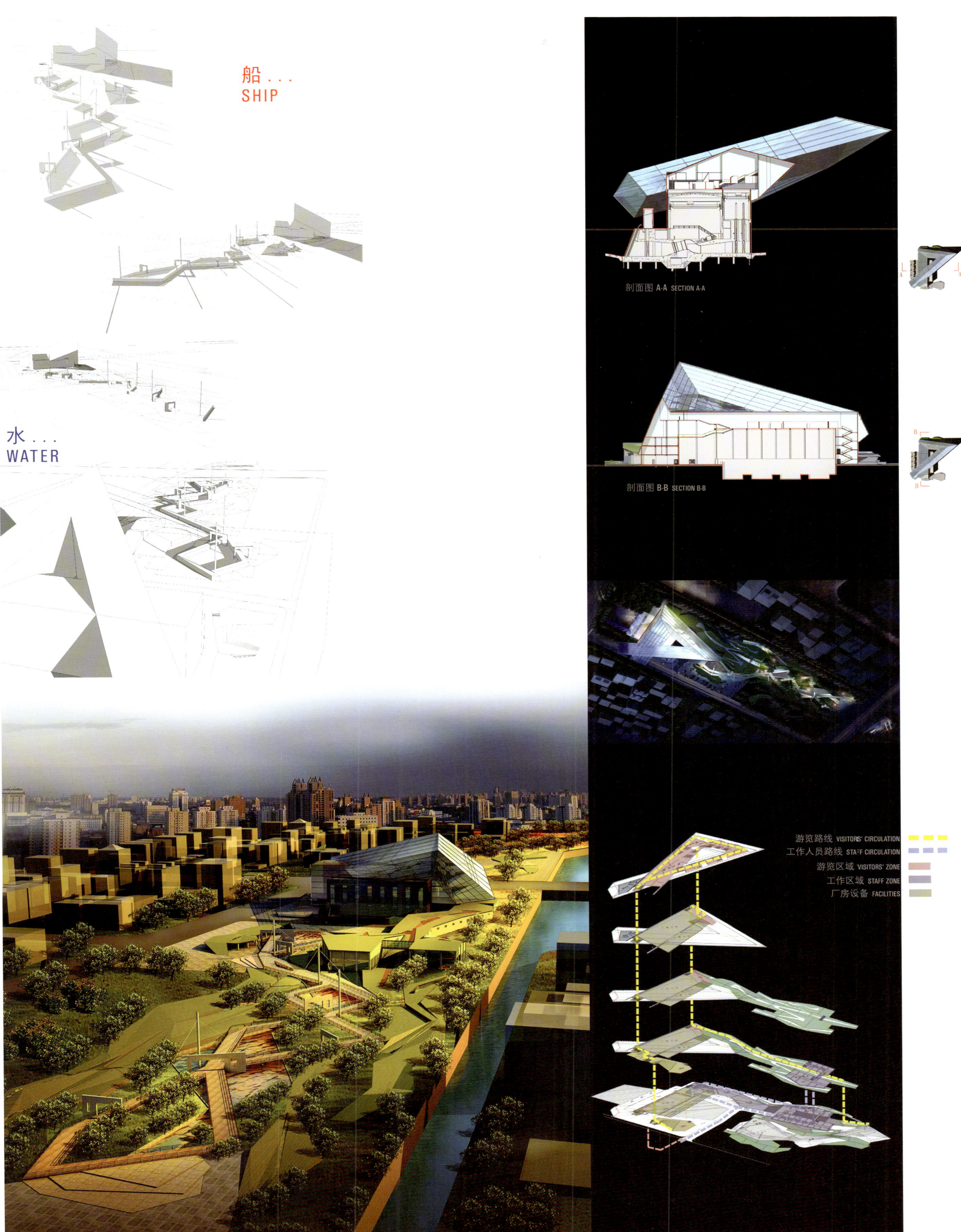
船...
SHIP
水...
WATER
剖面图 A-A SECTION A-A
剖面图 B-B SECTION B-B
游览路线 VISITORS' CIRCULATION
工作人员路线 STAFF CIRCULATION
游览区域 VISITORS' ZONE
工作区域 STAFF ZONE
厂房设备 FACILITIES

西湖博物馆 · 杭州

West Lake Museum · China

余健 Yu Jian · 曾勤 Zeng Qin · 黎冰 Li Bing · 王雷 Wang Lei · 沈济黄 Shen Jihuang · 沈金 Shen Jin · 裘涛 Qiu Tao · 刘晓梅 Liu Xiaomei · 卢德海 Lu Dehai · 曹志刚 Cao Zhigang

杭州西湖博物馆是为展现西湖的自然地理、人文历史而立，选址于西湖南线环湖景区，建筑用地约23889平方米，总建筑面积7920平方米，其中地上建筑面积1980平方米。建筑以考古发掘探沟和探方形态喻象探索西湖的历史，从而将大部分建筑设置于地下，南部以屋顶草坪缓坡的方式将建筑与西湖周边的绿化自然相接，建筑与环境一气呵成，含蓄而又不失个性。探沟的形态既契合了西湖边的西湖博物馆这一特定地点特定建筑的特点，同时也自然地引导了参观者的参观路线。设计尊重现有钱王祠路三世五王牌坊甬道的空间格局，在建筑北端以一道古典红墙实现了现代形态的西湖博物馆与传统形式的钱王祠之间的衔接与转换。

Hangzhou West Lake Museum is founded to present the physical geography and cultural history of West Lake. The Museum is located at a scenic area around the south line of West Lake, covering a site of around 23,889 sqm. The total construction area of the museum is 7,920 sqm and the above ground construction area is 1,980 sqm. The building compares a deep trench with the exploration of the history of West Lake; thus, most parts of the building are arranged underground. The building joins naturally with the surrounding plantation of West Lake in the form of a roof, lawn and gradual slope. The building merges perfectly with the surroundings, which are imponderable and unique. The shape of a deep trench not only is compatible with the characteristics of West Lake Museum shaping a particular building, but also guides the route of visitors in a natural way. The design preserves the spatial pattern of the existing memorial arches and paved paths of the period of the five emperors of the three generations of Emperor Qian's Temple, a classical red wall to the north which makes the connection and transition between the modern West Lake Museum and the traditional Emperor Qian's Temple.

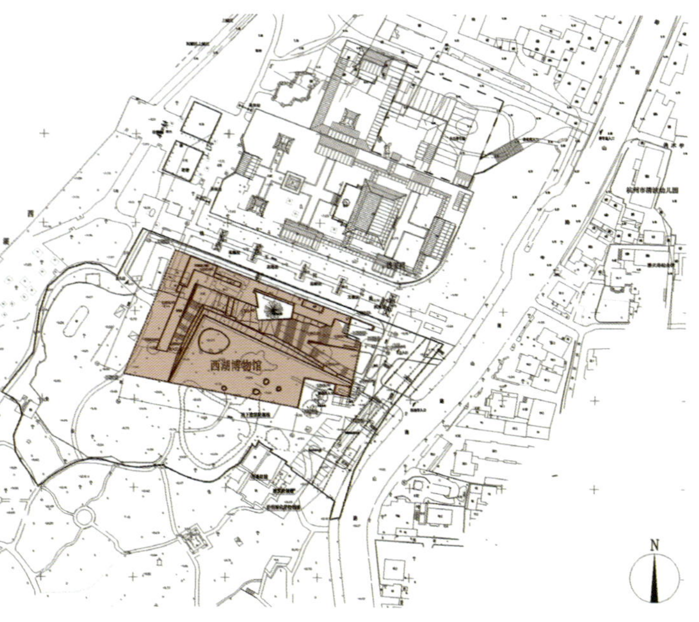

ZUCCEA (浙江大学建筑工程学院)
方案主创 project architect · 余健 Wang Zhu
项目负责 project managers · 曾勤 Tang Jun · 黎冰 Chen Zongyan
建筑设计 design architect · 王雷 Guo Wendong
设计指导 design director · 沈济黄 Shen Jihuang
结构工程 structural engineering · 沈金 Shen Jin · 裘涛 Qiu Tao
给排水工程 plumbing installation · 刘晓梅 Liu Xiaomei
电气工程 electrical installation · 卢德海 Lu Dehai
暖通工程 heating installation and ventilation · 曹志刚 Cao Zhigang
2006年浙江省优秀建筑设计一等奖
2008年全国优秀建筑设计一等奖
First Prize of 2006 Outstanding Architectural Design of Zhejiang Province
First Prize of 2008 National Outstanding Architectural Design

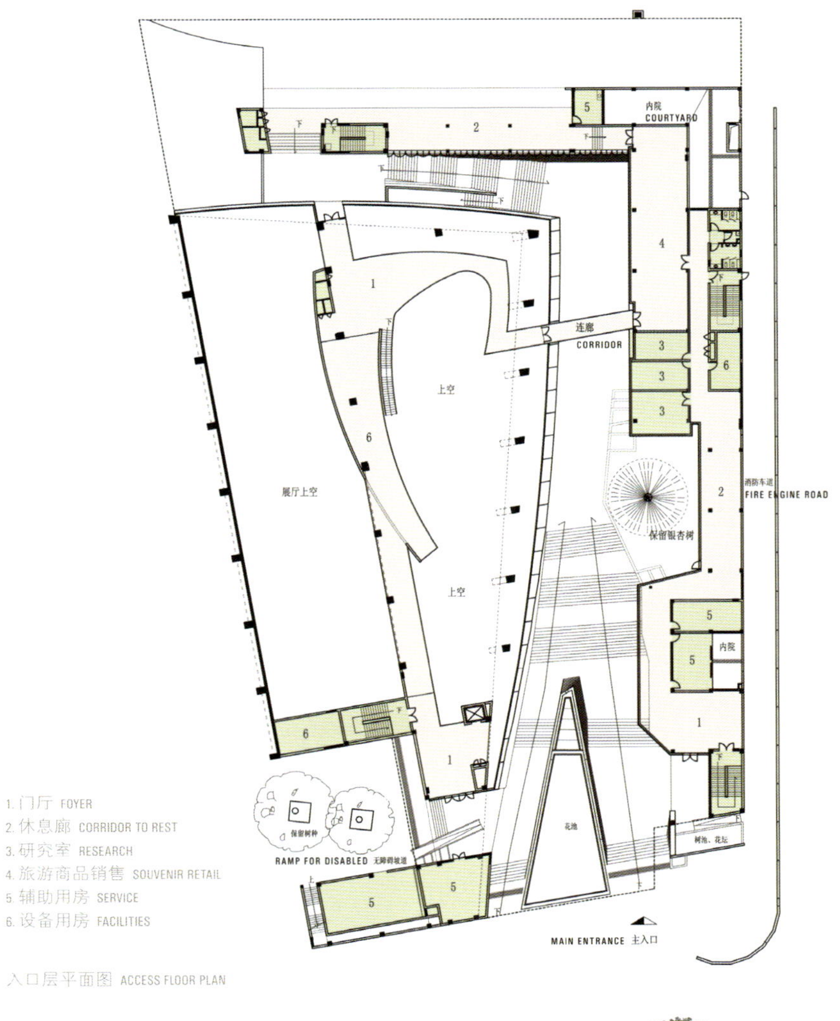

入口层平面图 ACCESS FLOOR PLAN

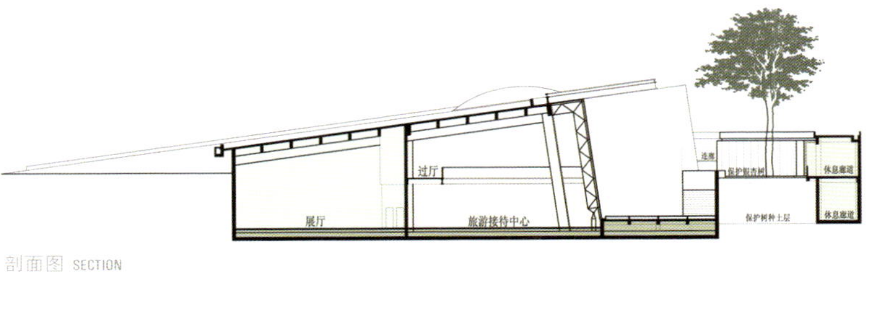

剖面图 SECTION

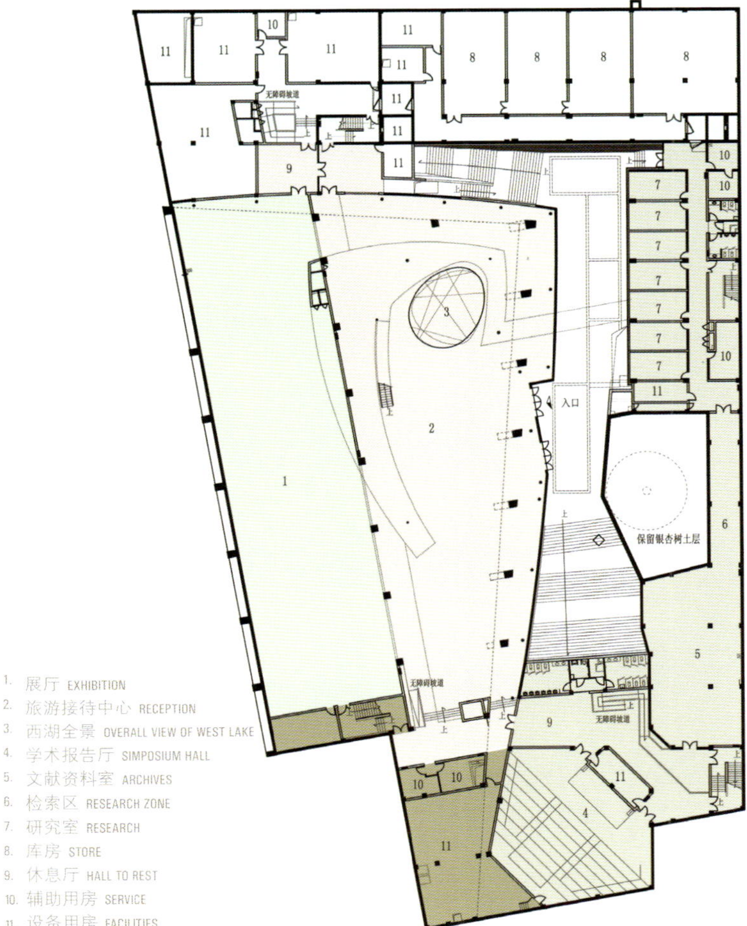

上层平面图 UPPER FLOOR PLAN

北京妫河建筑创意区

Beijing Gui River Creative Zone · China

竞标 · competition
北京妫河建筑意区
Beijing Gui River Creative Zone Yanqing

竞标类型 · competition type
一轮制邀标
A single stage with invitation

项目地点 · site area
延庆 · 北京 · 中国 Yanqing · Beijing · China

主办方 · promoter
北京市延庆县人民政府 People's Government of Yanqing County, Beijing Municipality
北京市建筑设计研究院 Beijing Institute of Architectural Design

日程安排 · schedule
招标 · Announcement 07.2009
评审结果 · Jury´s results 11.2009

获奖者 · awards

一等奖 · **first prize**
UNStudio (建筑师事务所)
Ben van Berkel (建筑师)

设计团队 design team · Caroline Bos, Markus van Aalderen, Yi Cheng Pan and Mieneke Dijkema, Ren Yee, Clarissa Alfrink, Megan Ng, Maud van Hees, Wendy van der Knijff, Jae Young Lee
城市发展顾问 advisor urban development · dcu, Pablo Vaggione
可持续发展顾问 advisor sustainability · Arup Shanghai

二等奖 · **second prize**
Atelier d'Architecture Christian de Portzamparc
(建筑师事务所)

入围 · **finalist**
AS. Architecture-Studio (建筑师事务所)

入围 · **finalist**
KCAP (建筑师事务所)
Han van den Born, Kees Christiaanse
Ruurd Gietema, Irma van Oort (建筑师)

北京妫河建筑创意区遵循延庆总体规划的要求，作为北京延庆设计创意产业园区一期项目，打造国际一流的集创作、培训、科研、成果展示、文化交流等全功能的创新型文化创意产业聚落。

依托延庆县的环境资源优势及其妫水河城市生态景观带，倡导生态、环保、节能、科技、创新等新理念，**主张贯穿全程的绿色设计、绿色建筑、绿色运营。**

Beijing Yanqing River Creative Zone follows the requirements of the master plan of Yanqing county, as the first stage of the Beijing Yanqing design and industry Park, which includes innovative cultural programs such as creative, training, research, exhibition and cultural exchanges facilities.

Considering the advantage of environmental resources in Yanqing county and its urban ecological Riverbed, the projects had to work on new ideas such as ecological and environmental protection, energy saving, scientific and technological innovation and **green design, green buildings throughout the whole process.**

北京妫河建筑创意区 · 延庆

Beijing Gui River Creative Zone · China

UNStudio (建筑师事务所)

一等奖 · First Prize

大自然中的都市风

“大自然中的都市风”这一主题是指保护延庆自然景观的同时也充分打造都市生活。该设计包含了两个要素：绿化带和都市建筑体。都市建筑体由两个相互交织的地带组成——“创意产业”地带和“配套设施”地带，这两个地带风格迥异却联系紧密，共同打造了一系列开放性空间，如内部街道、庭院、广场入口、公共街区等。在建筑公园中，分散的工作室和展馆均按照中国传统的园林格局进行布置，达到了“一步一景”的布局效果。

URBANITY IN NATURE

The theme of Urbanity in Nature refers to the preservation of the natural qualities of the Yanqing site, whilst accommodating a certain essential level of urban city life density. The design encompasses compression on a dual scale; the greenbelt and the urban fabric. The urban fabric of the edge is formed by the interweaving of two distinct yet highly interconnected strips - the “Creative Industry Based” strip and the “Supporting Facilities” strip, and consequently creates a series of open spaces ranging from internal streets to courtyards and plaza entrances to public squares. In the creation of the Architectural Park, the scattered ateliers and pavilions are organized on the basis of the Chinese garden's emphasis of “一步一景” (“One View Per Step”).

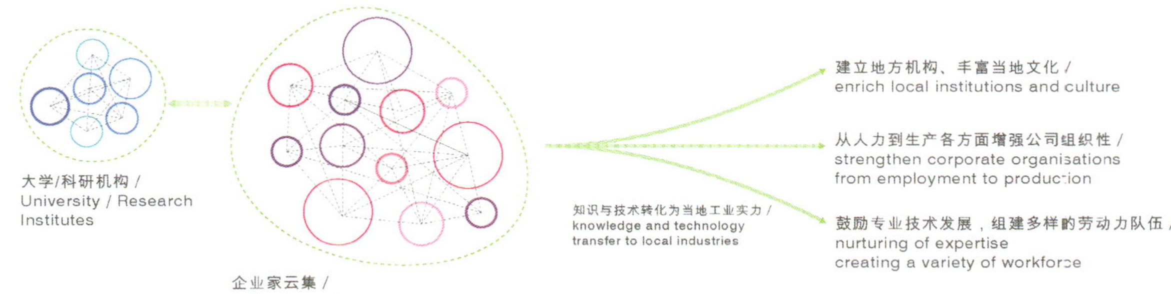

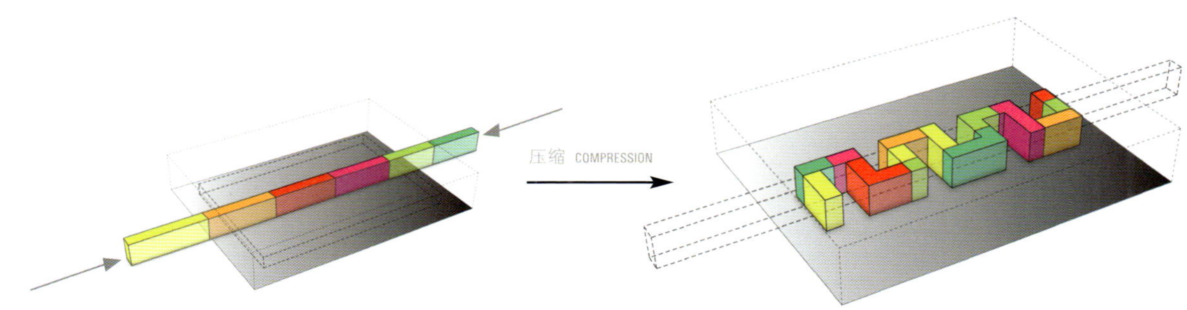

底层平面图 GROUND FLOOR PLAN

教育培训 EDUCATION+TRAINING

会议厅 LECTURE HALL
教室 CLASSROOMS
设计工作室 DESIGN STUDIOS
画廊 GALLERY
配套服务设施 SUPORTING SERVICES FACILITIES

创意工作室 CREATIVE STUDIOS

中小型 MEDIUM AND SMALL SIZED
研发工作室 R&D STUDIOS
建筑大师工作室 MASTER ARCHITECTURE STUDIO
娱乐 RECREATION
配套服务设施 SUPORTING SERVICES FACILITIES

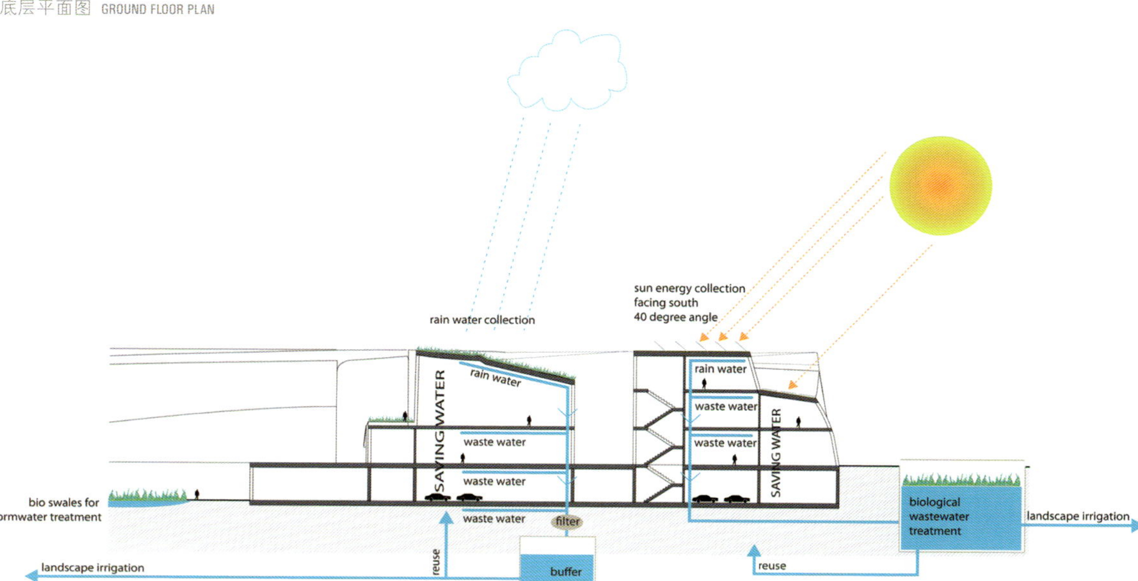

水系统及太阳能 WATER SYSTEM & SUN ENERGY

展览交流 EXHIBITION+EXCHANGE

展览交流 EXHIBITION+EXCHANGE
画廊 GALLERY

一分钟 1 MINUTE
两分钟 2 MINUTES
三分钟 3 MINUTES

底层平面布局 GROUND FLOOR PROGRAM

主题建筑和酒店 THEME BUILDING+HOTEL

酒店 HOTEL

- 酒店 HOTEL FACILITIES
- 酒店 HOTEL FACILITIES
- 咖啡及艺术沙龙 CAFE & ART SALON

主题建筑 THEME BUILDING

- 公寓 HOUSING
- 辅助 SUPPORTING SER
- 博物馆 MUSEUM
- 研发工作室 R&D STUDIOS
- 办公室 OFFICES
- 辅助 SUPPORTING SER

二层平面布局 FIRST FLOOR PROGRAM

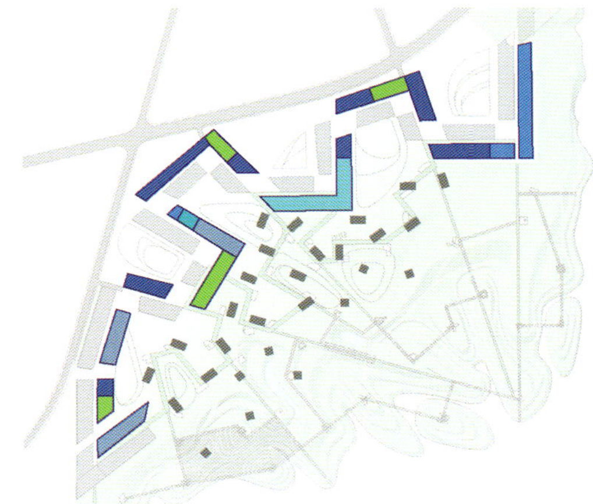

辅助 SUPPORTING SER

- 创意工作室 CREATIVE STUDIOS
- 配套服务设施 SUPORTING SERVICES FACILITIES
- 画廊 GALLERY
- 公寓 HOUSING/APARTMENTS

北京妫河建筑创意区 · 延庆

Beijing Gui River Creative Zone · China

Atelier d'Architecture Christian de Portzamparc (建筑师事务所)

二等奖 · Second Prize

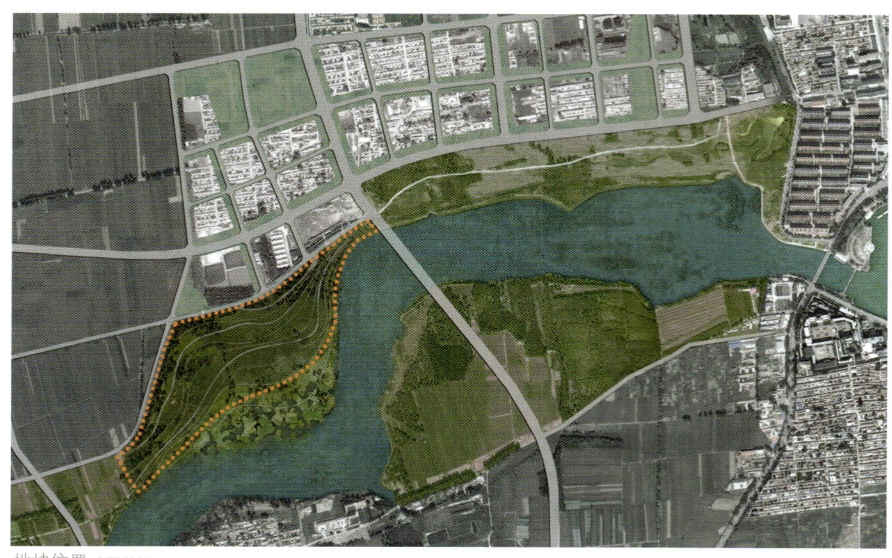

地块位置 SITE PLAN

- 酒店 HOTEL
- 教育与培训中心 TRANING CENTRE
- 创意工作室 STUDIOS CREATIVOS
- 配套服务设施 SERVICE FACILITIES
- 会展交流 EXHIBITIONS
- 博物馆 MUSEUM

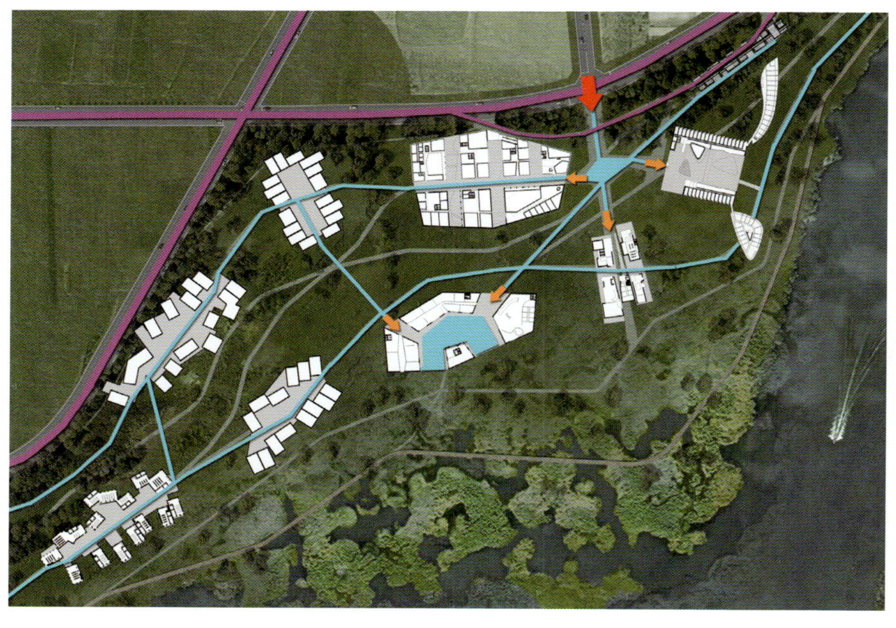

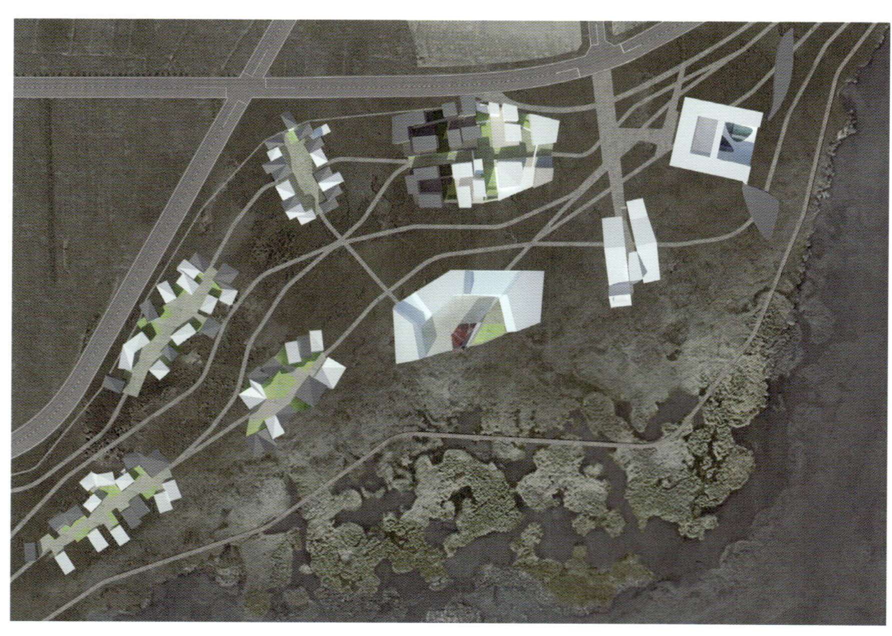

酒店 HOTEL

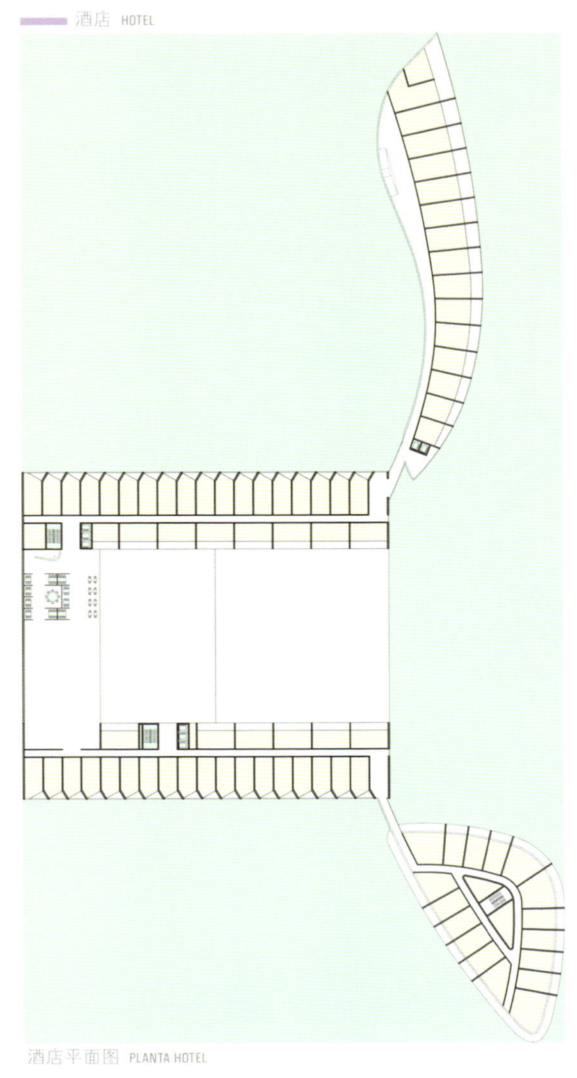

酒店平面图 PLANTA HOTEL

建筑学院 ARCHITECTURE SCHOOL

建筑工作室 ARCHITECT'S STUDIO

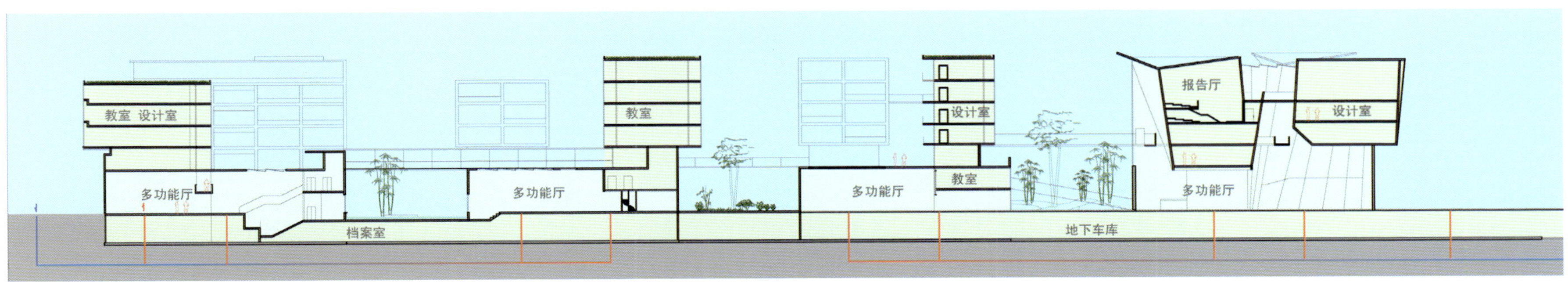

博物馆 MUSEUM

北京妫河建筑创意区 · 延庆

Beijing Gui River Creative Zone · China

入围 · Finalist

开放性设计

创意综合体是一个开放性的流动空间，空间内可进行信息、知识及经验交流。该项目的三项基本要素是高山、建筑和水。创意村包括两个基本功能：工作—生活空间和公共服务空间。公园具备两大功能：一个是集酒店、展览交流和教育设施为一体的城市功能，另一个则是其本质功能，如创意工作室和展廊。公园内有许多与延庆这个城市规模相适应的配套设施，这些设施，如餐厅、商店、酒吧、图书馆等，将公园的这两大功能巧妙地融为一体。

OPEN DESIGN

The creative complex is an open flowing space, in which there is an exchange of information, knowledge and experience. The three basic elements of this project are the hills, the buildings and the water. The creative village consists of two basic functions: working-living spaces and public service spaces. The park functions are divided into 2 parts: one part with city functions, such as a hotel, exhibition exchange, educational facilities; another with natural functions such as a creation studio and an exhibition passage. There are facilities in relationship with Yanqing city's scale to connect these 2 functions: a restaurant, shop, bar, library...

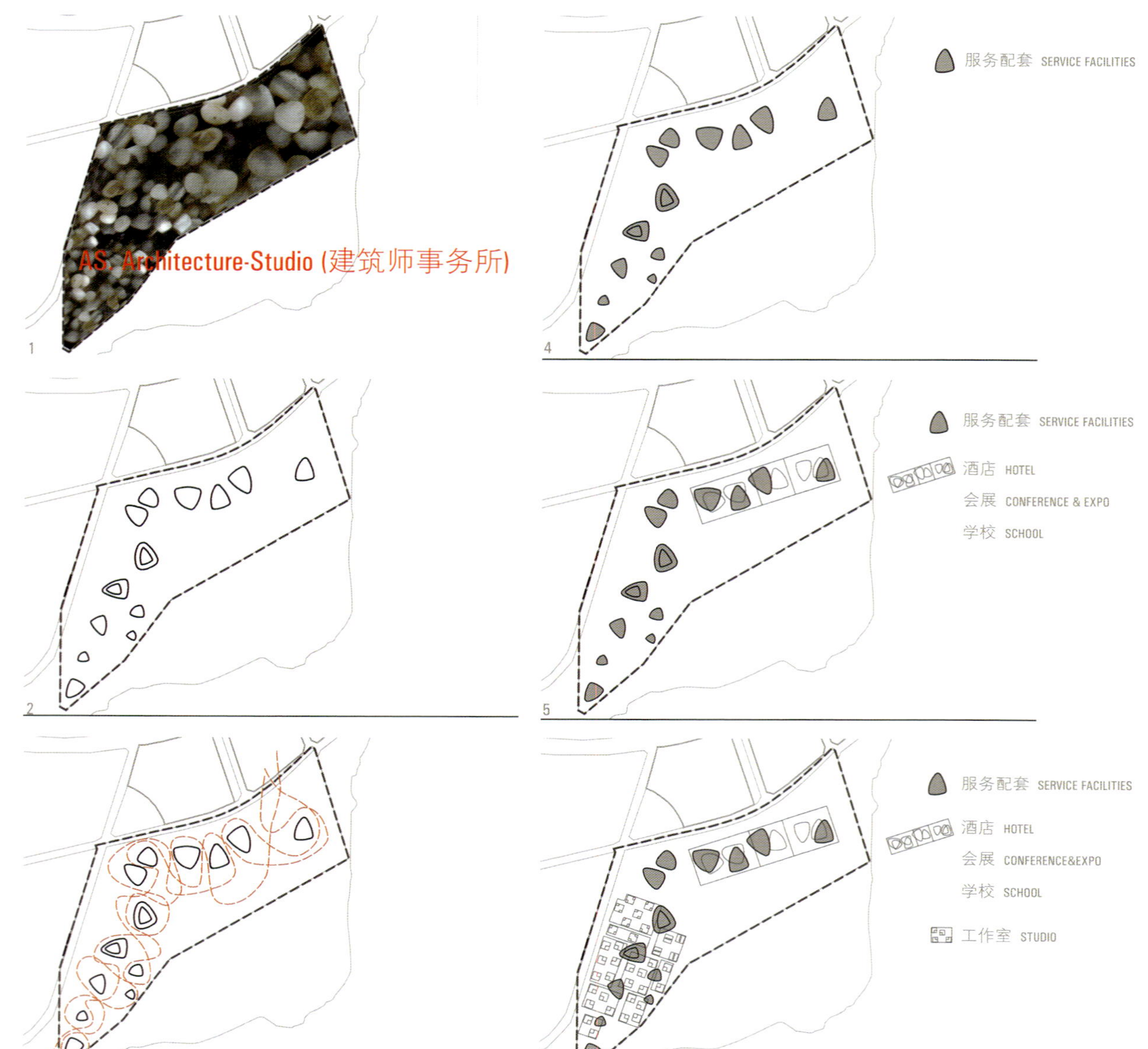

AS. Architecture-Studio (建筑师事务所)

规划结构及功能分析图 URBAN STRUCTURE AND FUNCTIONS PLAN

总平面图 MASTER PLAN

总剖面图 MAIN SECTION

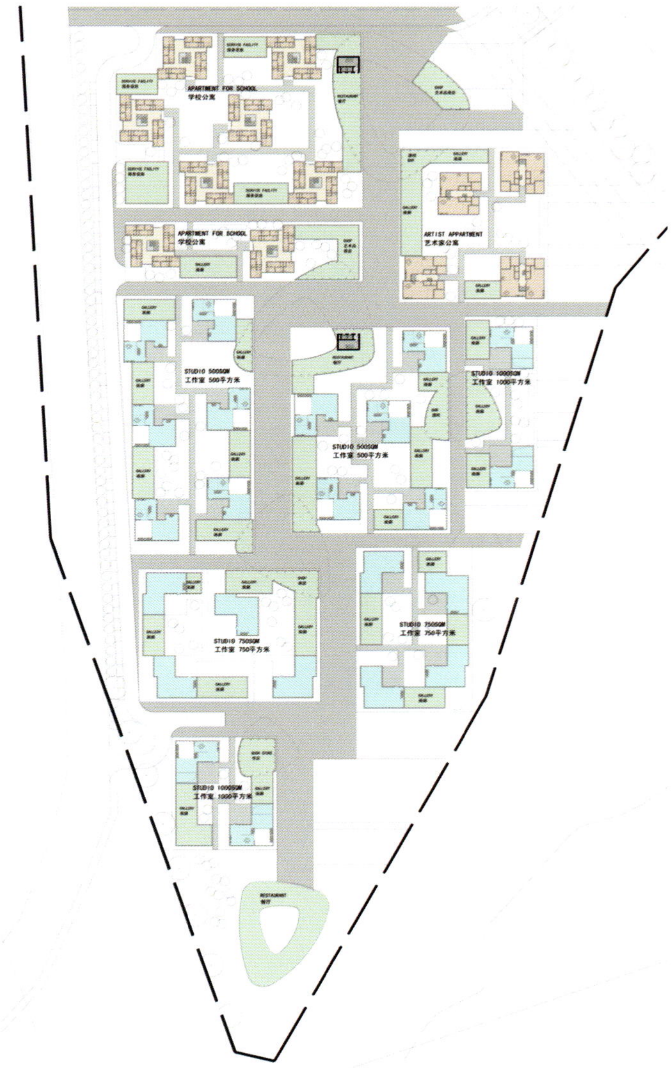

创意村底层平面图 GROUND FLOOR PLAN CREATIVE VILLAGE

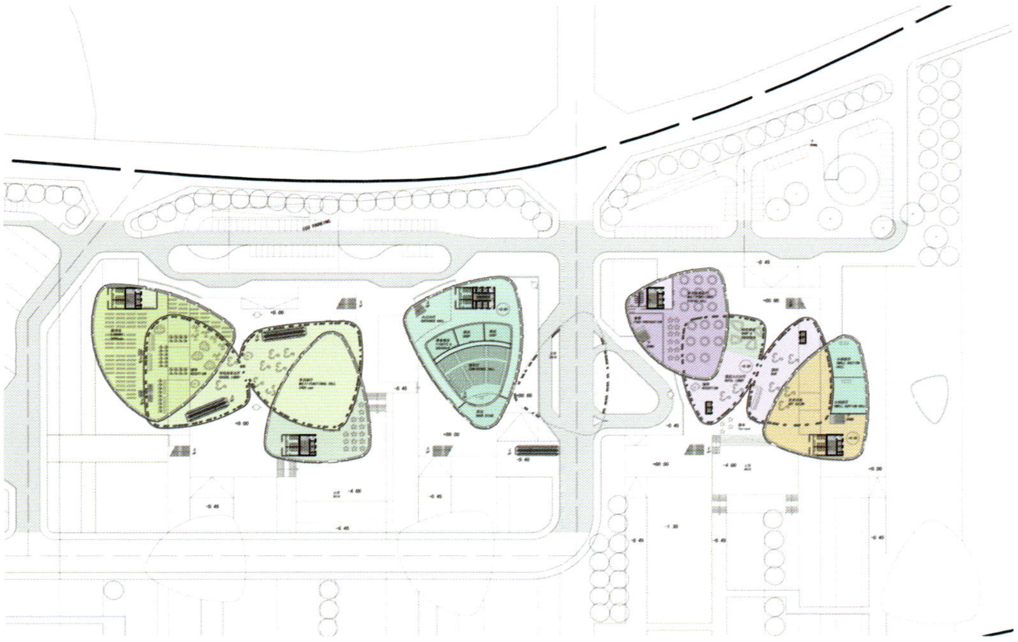

主要建筑底层平面图 GROUND FLOOR PLAN MEGACITY

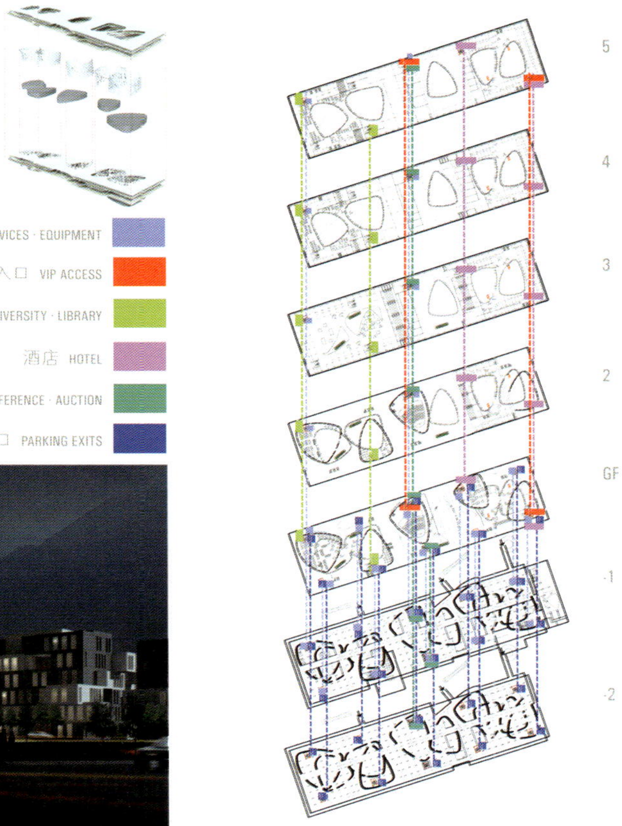
服务及设备 SERVICES · EQUIPMENT
嘉宾入口 VIP ACCESS
大学及图书馆 UNIVERSITY · LIBRARY
酒店 HOTEL
会展及拍卖会 CONFERENCE · AUCTION
停车场出口 PARKING EXITS
5
4
3
2
GF
-1
-2

建筑功能及流线分析 FUNCTION AND ACCESS ANALYSIS

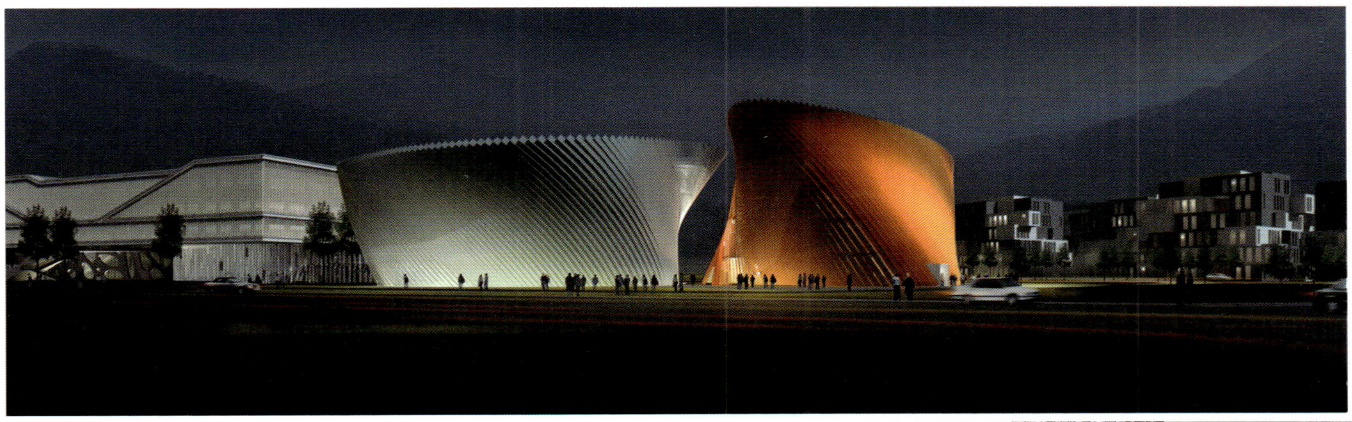

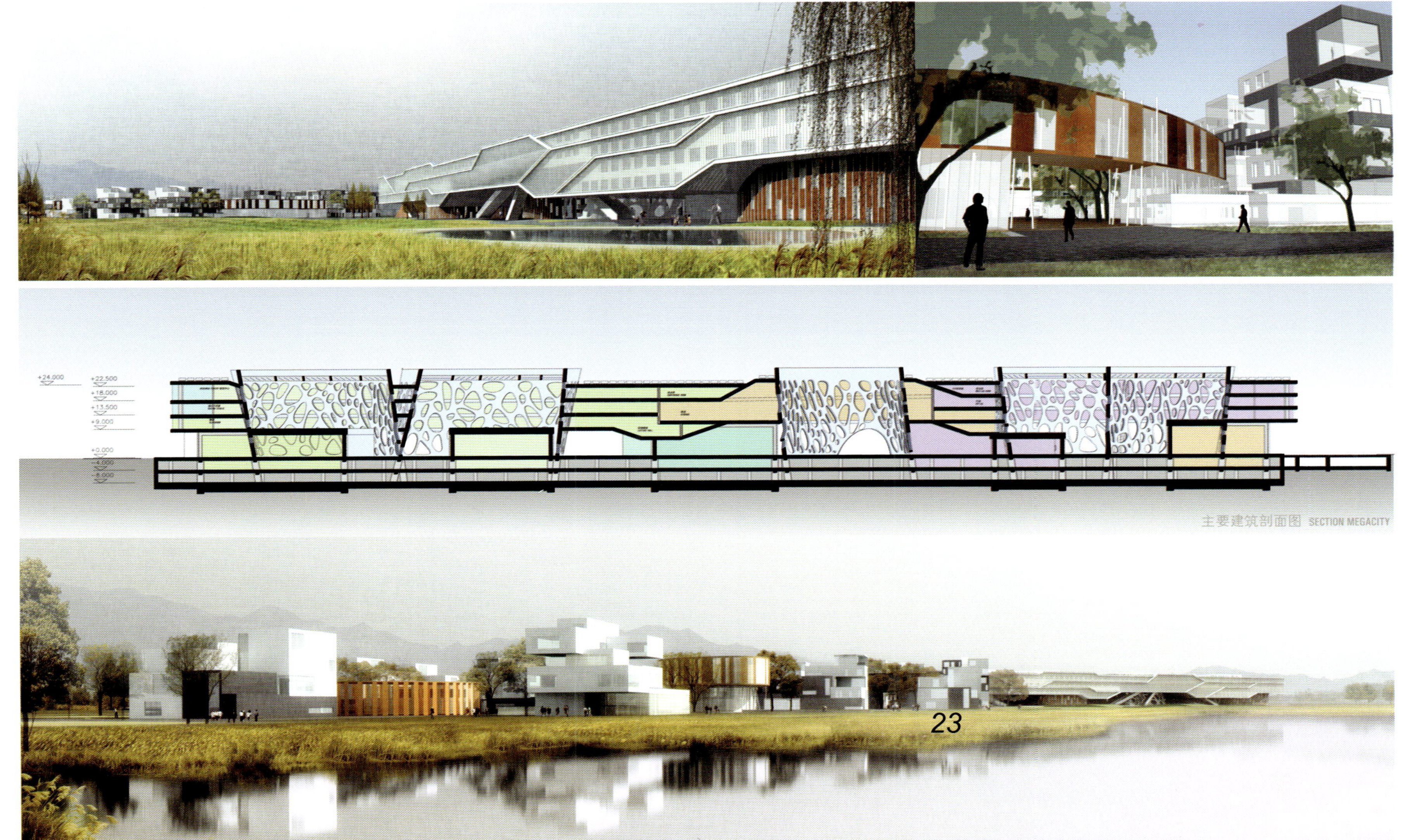

主要建筑剖面图 SECTION MEGACITY

北京妫河建筑创意区·延庆

Beijing Gui River Creative Zone · China

KCAP (建筑师事务所)

入围 · Finalist

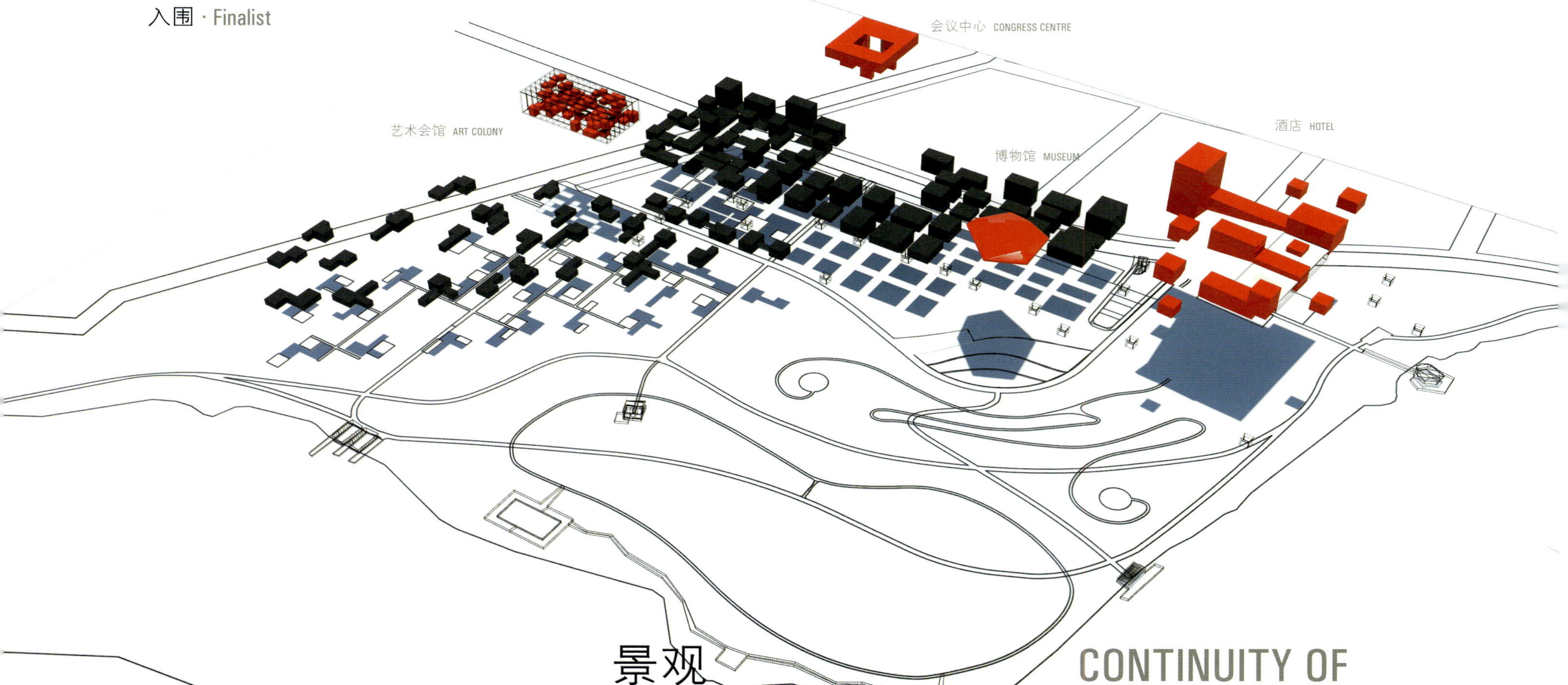

会议中心 CONGRESS CENTRE

艺术会馆 ART COLONY

博物馆 MUSEUM

酒店 HOTEL

景观连续体

CONTINUITY OF THE LANDSCAPE

三大区域沿东西方向铺展开来：一是生产区，环境一流；二是交流区，屹立于公共空间上的悬臂式建筑，能在炎热的天气里遮阴蔽日；三是休息区，供人们平静、放松和休息。四幢建筑都各有特点：规划区西侧的“艺术群落”、中部的会议中心和博物馆以及东侧的酒店。这些建筑体加速了城市各项活动的开展。

Three main areas in west-eastern direction are envisioned: A Production zone providing perfect working conditions; An Exchange zone with cantilevered buildings on the public space and creating shelter from the hot weather conditions; A Contemplative zone of peace, relaxation and space to rest. Four buildings are designated as key carriers of identity: An 'Art Colony' at the western edge of the site, a Congress Centre and a Museum in the middle and the Hotel at the eastern edge of the planning area. Those structures are envisioned as the catalysts of the urban activity.

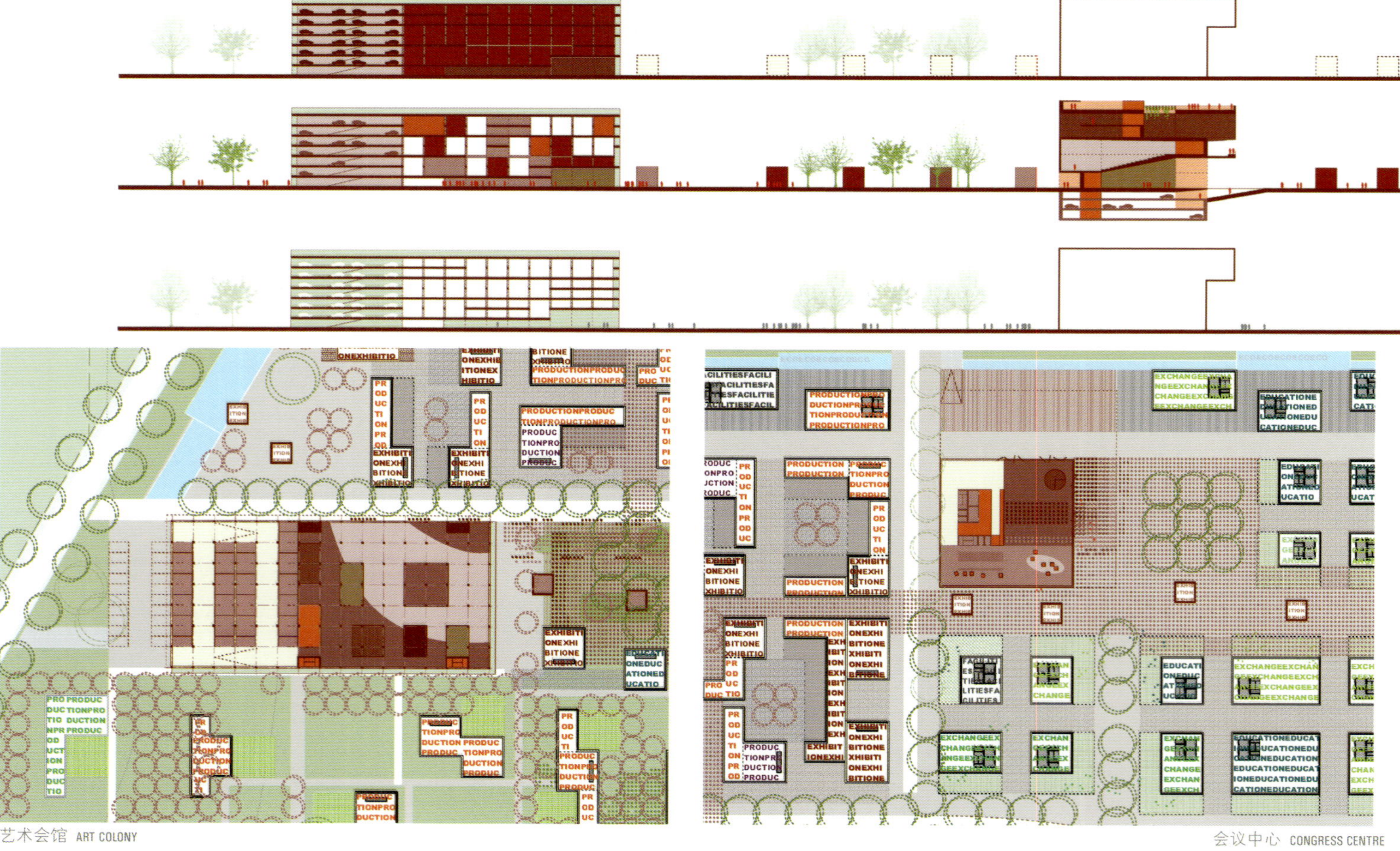

艺术会馆 ART COLONY

会议中心 CONGRESS CENTRE

地块位置 SITE PLAN

底层总平面图 GROUND FLOOR PLAN

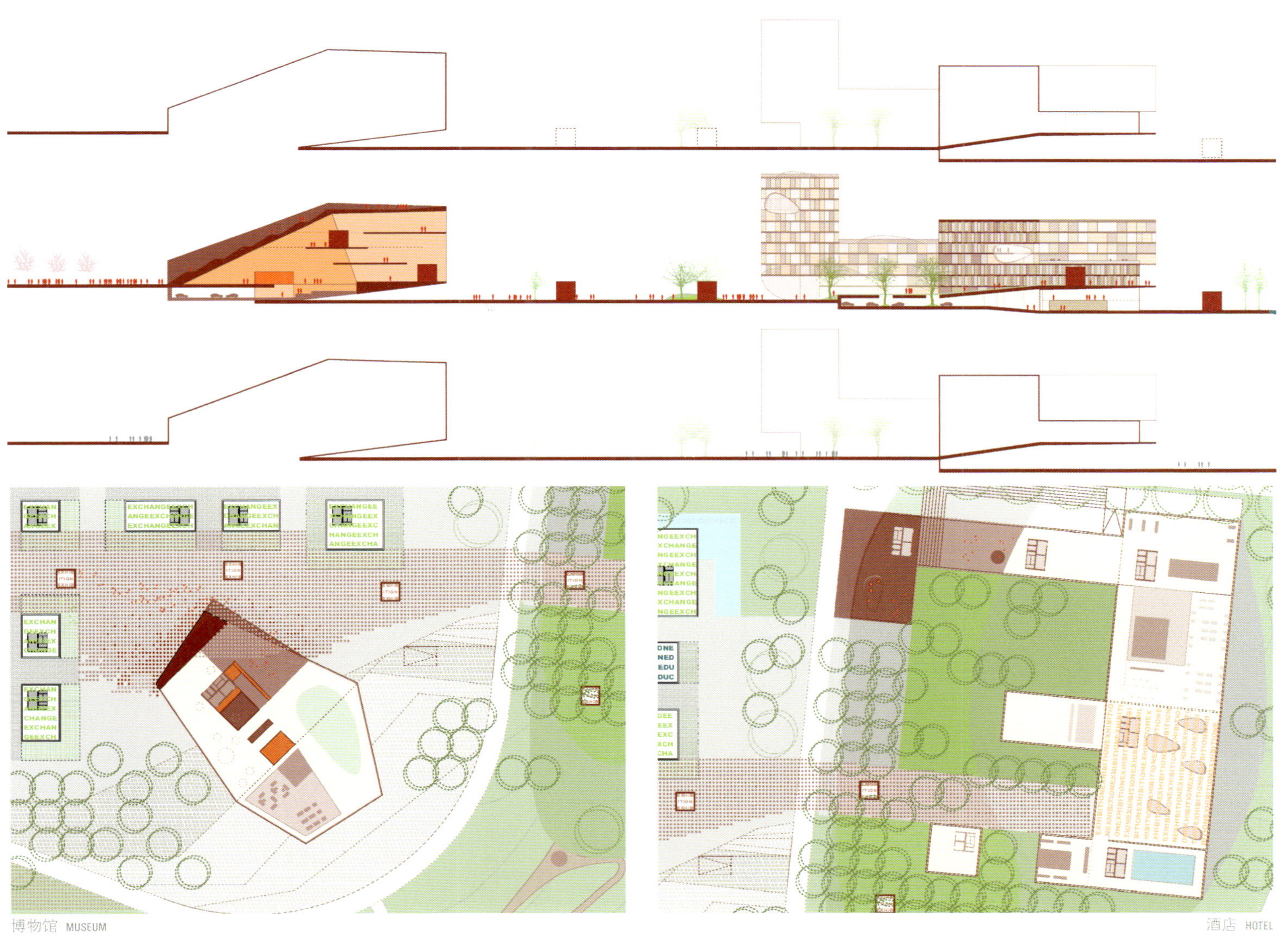

博物馆 MUSEUM

酒店 HOTEL

克里普塔纳乡村剧院礼堂 · 雷阿尔城
Campo de Criptana Theatre-Auditorium · Spain

竞标 · competition
克里普塔纳乡村剧院礼堂 · 雷阿尔城
Campo de Criptana Theatre-Auditorium · Spain

竞标类型 · competition type
一轮制概念公开竞标
open, one-stage design competition

项目地点 · site area
雷阿尔城 · 西班牙 Ciudad Real · Spain

主办方 · promoter
克里普塔纳村政府及卡斯蒂利亚拉芒查雷阿尔城区官方建筑学院
Ayuntamiento de Campo de Criptana con la colaboración del Colegio Oficial de Arquitectos de Castilla-La Mancha demarcación de Ciudad Real

日程安排 · schedule
招标 · Announcement 01.2008
评审结果 · Jury´s results 09.2008

评审团 · jury

主席 · presidente:
Santiago Lucas-Torres López-Casero
委员 · vocales:
Teresa Rodríguez-Rey Expósito
Francisco Javier Bernalte Patón
Angel Quintanar Cortés
Ignacio Díaz-Ropero Cruz
Luis Cobos Pavón
Manuel Manzaneque Díaz-Hellín
秘书 · secretario:
Antonio Cedenilla Díaz-Hellín

获奖者 · awards

中标 · winner
CAMACHO + MACIÁ ARQUITECTOS (建筑师事务所)
Javier Camacho · María Eugenia Maciá

合作 (c) María Navascués · Christine Müller-Hillebrand · Juan Santana · Alejandro Postigo

入围 · finalist
FRECHILLA&LÓPEZ-PELAEZ ARQUITECTOS (建筑师事务所)

建筑师 arquitectos: Javier Frechilla · Lorenzo Gil · Luis Gil · José Silva · Carmen Herrero
José Manuel López-Peláez · Luis Martinez · Diego Palomares · Adolfo Navarro
arquitectos técnicos: Emilio Rodríguez · Fernando Rilova

参标 · proposal
PANCORBO ARQUITECTOS (建筑师事务所)
Luis Pancorbo · Ines Maria Martin · Jose de Villar

参标 · proposal
EVA ISZORO (建筑师)

因为缺乏塞万提斯剧院此类的资源，所以雷阿尔城的目标是为本市建造标志性建筑，用来举办音乐、戏剧文化活动。选用这片土地为建造地，是因为看中该地的**战略性区位**和开放式环境。

The goal is to provide the municipality with a landmark building which can house the cultural activities related with music and theatre, due to the lack of resources of the existing Cervantes Theatre. This plot has been selected due to its **strategic location** and its open environment.

克里普塔纳乡村剧院礼堂·雷阿尔城

Campo de Criptana Theatre-Auditorium · Spain

一等奖 · First Prize

endecaciclos (竞标代码)

Camacho + Maciá arquitectos (建筑师事务所)

Javier Camacho · María Eugenia Maciá (建筑师)

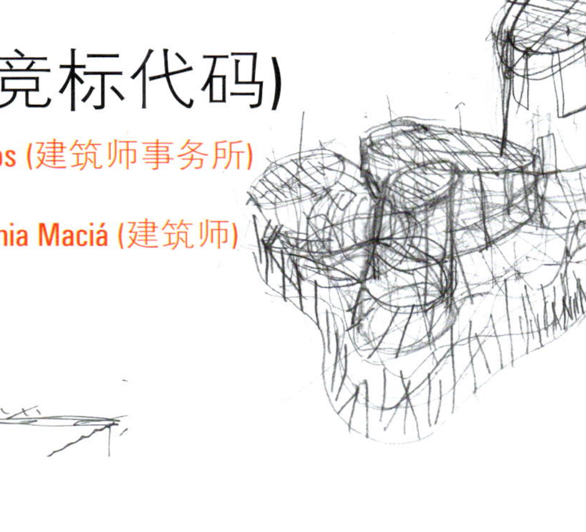

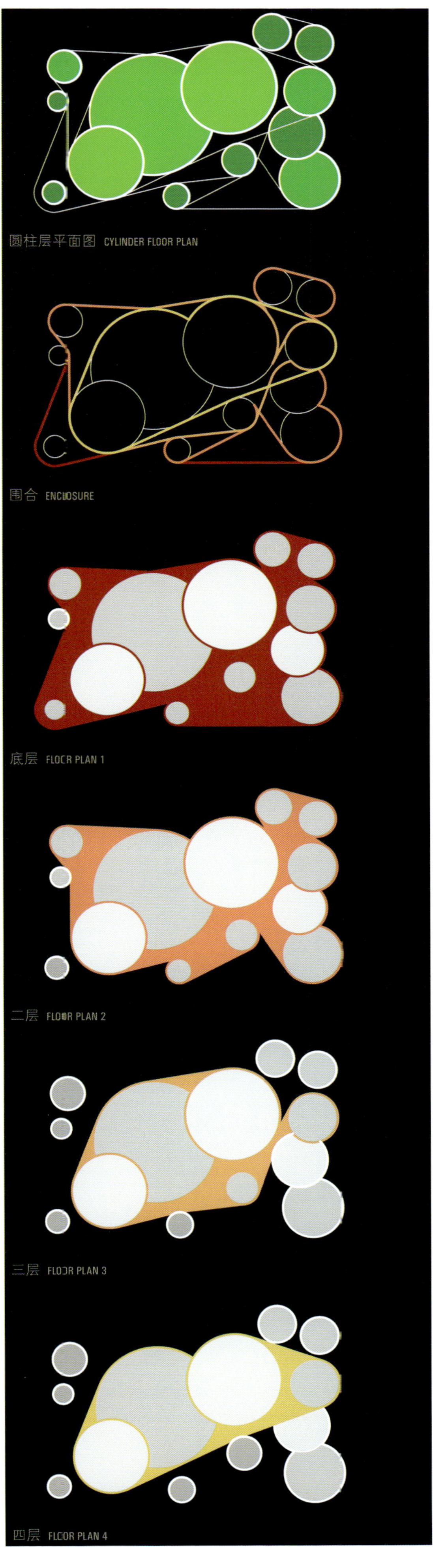

圆柱

CYLINDERS

我们希望调整礼堂（黑盒子）的容量大小以适应周围建筑的规模，并决定基本上取消所有平面图和主要楼层规划图，并完全遮挡黑盒子。这样，建筑的外观才能得以凸显。理想的正式法则基于克里普塔纳村的“磨坊”的单一建筑类型而建立，其圆柱表面进行不同的相互作用，形成了圆柱内、外部单一空间。

We look to adapt the big volume of the auditorium (the black box) to the scale of the rest of the buildings of the surroundings. We decide to almost bury the ground and main floor plan and to radically bury the black box. Thus the façade breaks up. The desirable formal laws base on the singular typology of "The Mills" of Campo de Criptana, through cylindrical surfaces which produce different interactions resulting singular spaces inside and outside the cylinders.

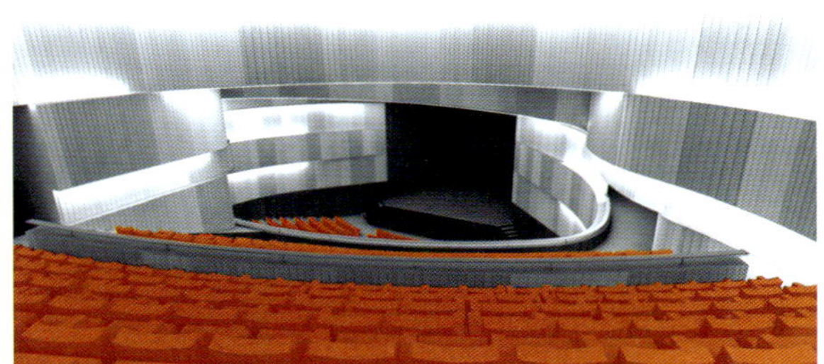

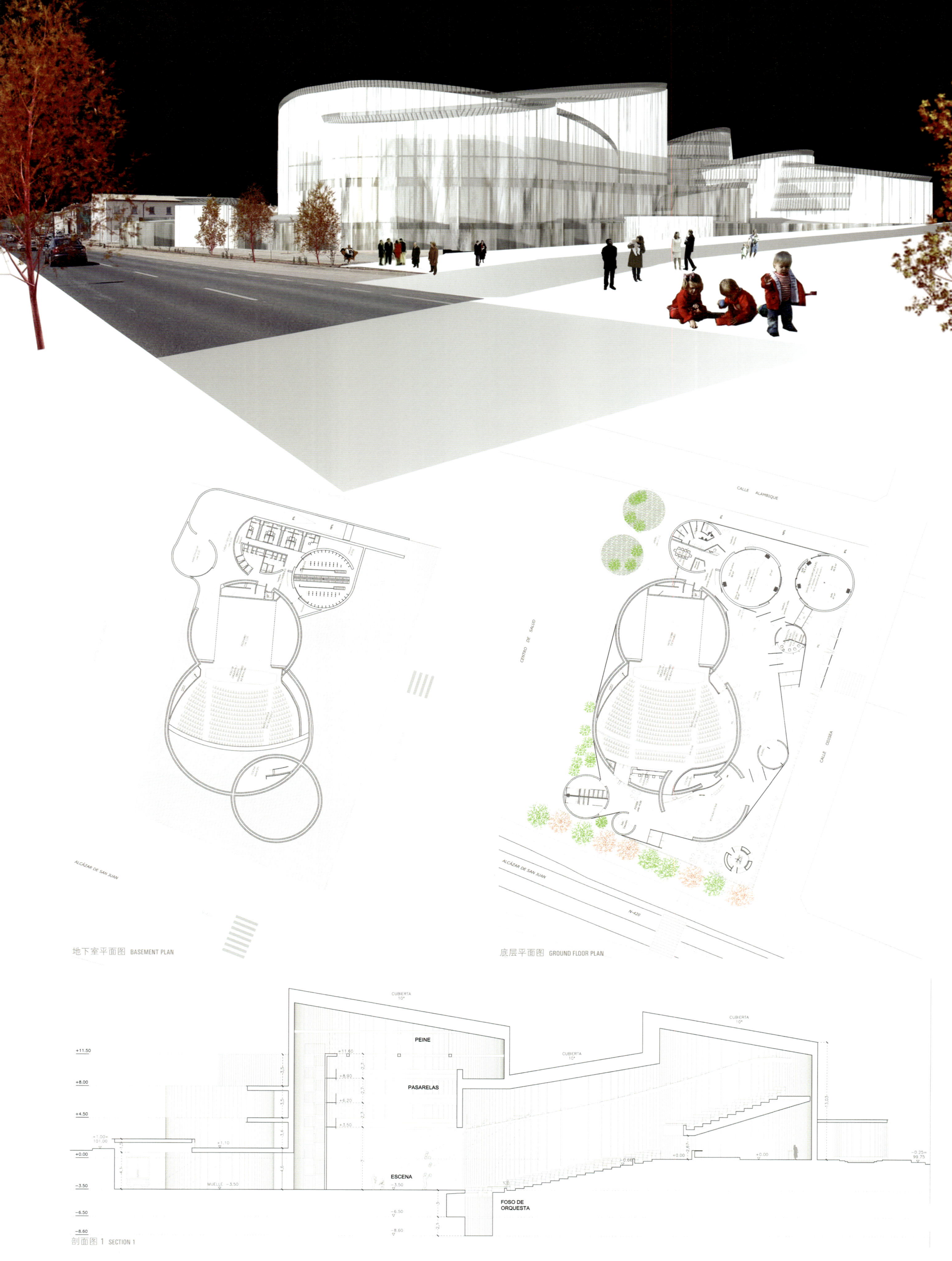

地下室平面图 BASEMENT PLAN

底层平面图 GROUND FLOOR PLAN

剖面图 1 SECTION 1

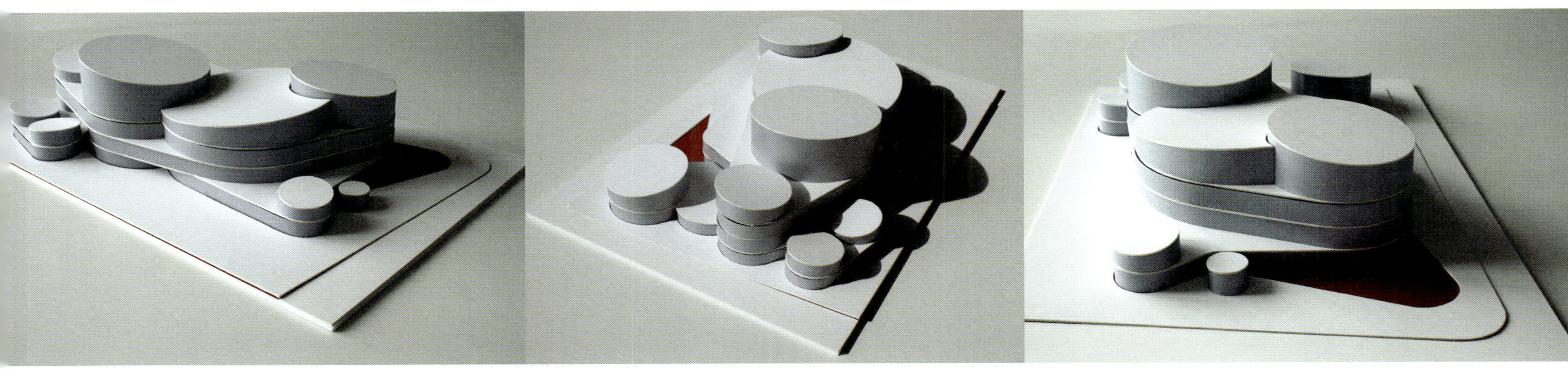

二层平面图 FLOOR PLAN 1

顶层平面图 ROOF PLAN

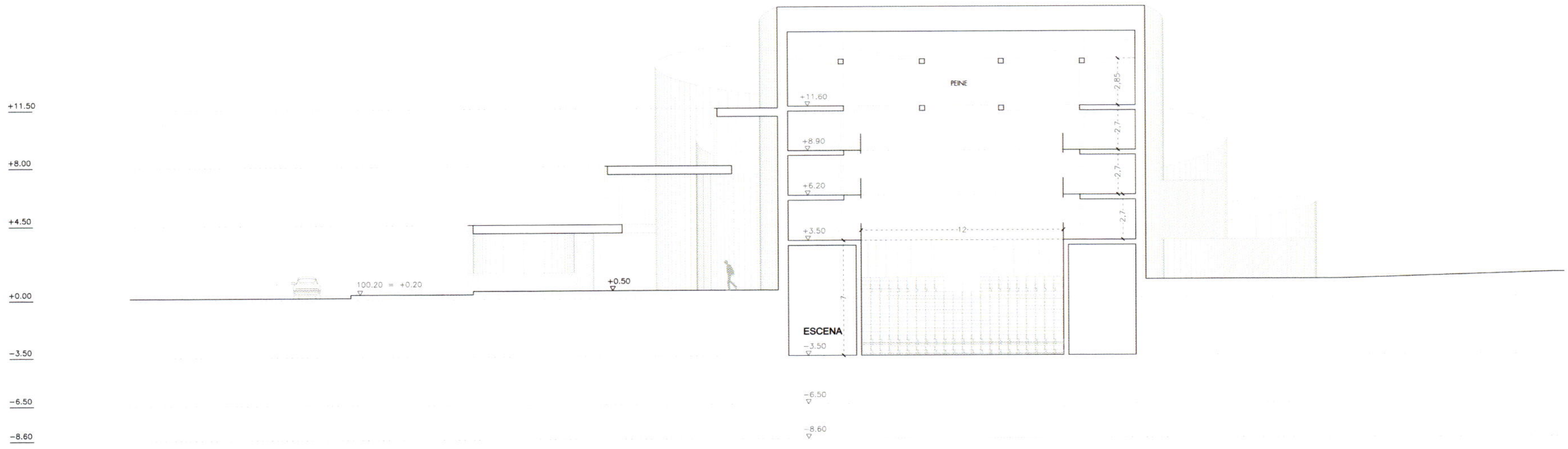

剖面图 2 SECTION 2

克里普塔纳乡村剧院礼堂 · 雷阿尔城

Campo de Criptana Theatre-Auditorium · Spain

入围 · Finalist

arara (竞标代码)

Frechilla-López Pelaez arquitectos (建筑师事务所)

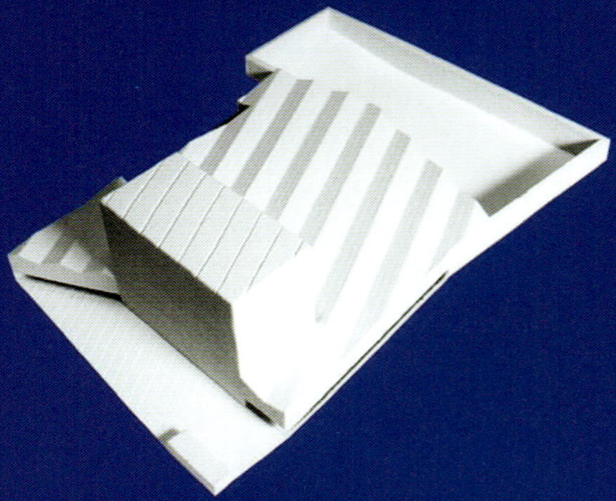

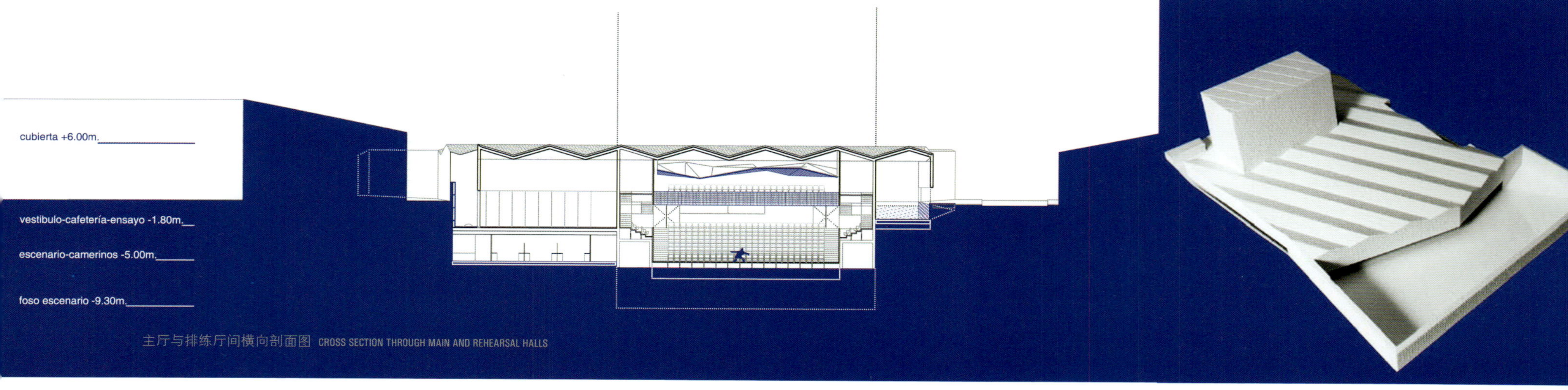

主厅与排练厅间横向剖面图 CROSS SECTION THROUGH MAIN AND REHEARSAL HALLS

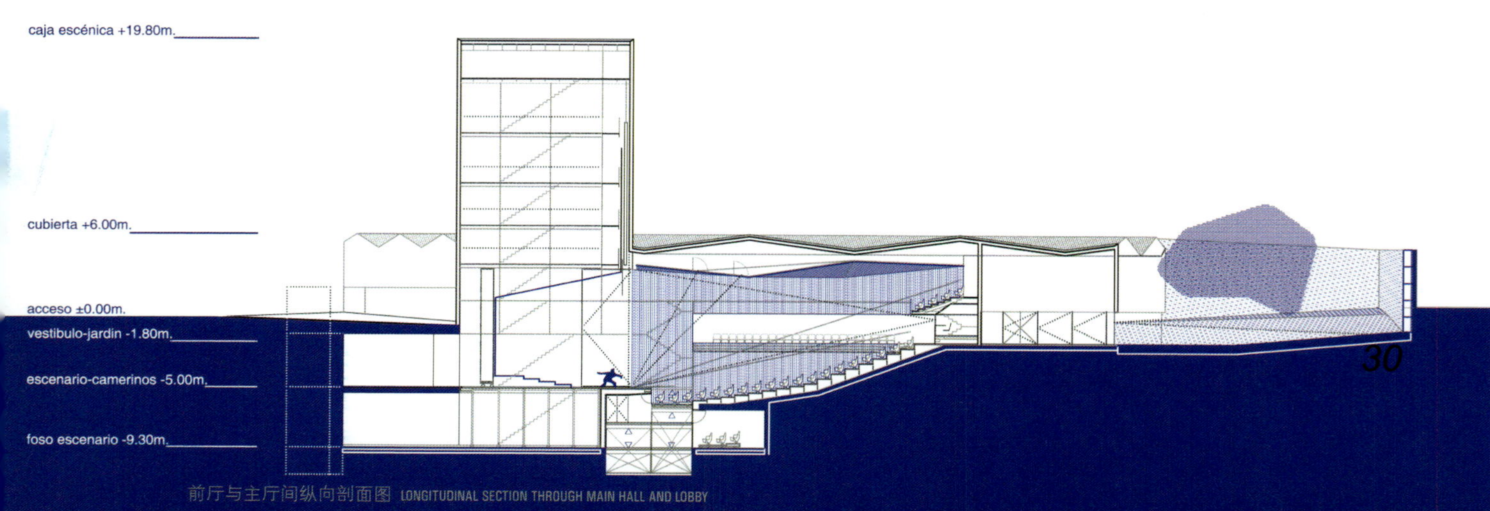

前厅与主厅间纵向剖面图 LONGITUDINAL SECTION THROUGH MAIN HALL AND LOBBY

城市周边环境

黑盒子“接近”工业园区，远离住宅区。实际上在整个住宅区内，除了黑盒子部分能凸显外，该建筑其他部分均被遮挡，因此并不显得突出。该建筑的6米平顶、墙壁和架构均采用建筑混凝土以及基于不同质地而选用的特殊建筑用模子材料。

INSIDE URBAN SURROUNDINGS

The volume of the black box "approaches" to the industrial park and moves away from the residential area. Because of the fact that the building is buried except for the black box volume, the rest of the building does not stand out from the general height of the residential city; six metres. Tiled roof, walls and structure made up of architectural concrete with special formwork depending on texture.

平面图 +1.60 FLOOR PLAN LEVEL +1.60

平面图 +0.00/-1.80 FLOOR PLAN LEVEL +0.00/-1.80

剧院礼堂多功能厅
ALTERNATIVE PROGRAMS TO THE THEATRE AUDITORIUM

排练厅·花园入口
REHEARSAL ROOMS · ACCESS THROUGH GARDEN

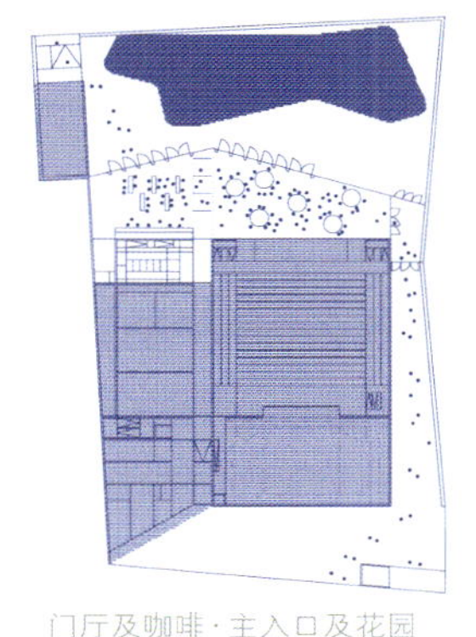

门厅及咖啡·主入口及花园
LOBBY AND CAFE · MAIN ACCESS AND GARDEN

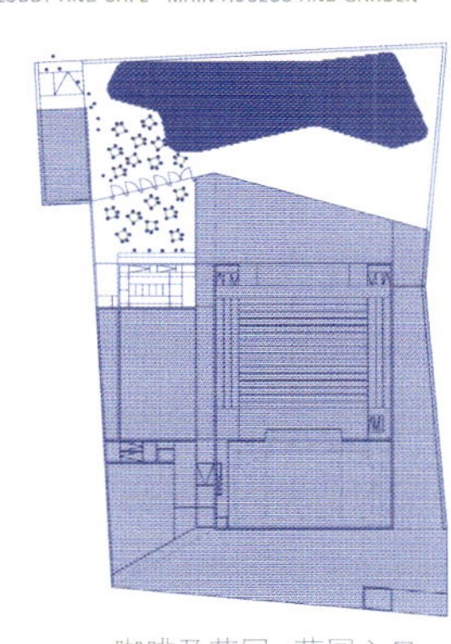

咖啡及花园·花园入口
CAFE AND GARDEN · ACCESS THROUGH GARDEN

平面图 -5.00 FLOOR PLAN LEVEL -5.00

克里普塔纳乡村剧院礼堂 · 雷阿尔城

Campo de Criptana Theatre-Auditorium · Spain

参标 · Proposal

t.a.c.c. (竞标代码)

Pancorbo arquitectos (建筑师事务所)

Luis Pancorbo · Ines Maria Martin · Jose de Villar (建筑师)

地块位置 SITE PLAN

寻找广告

我们正在寻找一个创意，使得该建筑既与住宅区的规模相符，同时作为一个城市图标，又不被遮挡，从路上即可见。我们构想建造一个6.50米的地基，这样能够与周围的建筑规模完美地融合。受拉芒查地区传统住宅的启发，我们选用一个暗含的白色元素，让其包裹整个庭院。黑盒子位于地基之上，面朝马路，在蓝色聚碳酸酯移动板的包围中，具有一定的标志性和代表性。

LOOKS FOR ADVERTISING

We look for an idea which is capable of adapting to the scale of the residential area and at the same time visible from the road as an urban icon. We propose a 6.50m base which perfectly adapts to the sixe of the surrounding buildings. A white and blind element which encloses the courtyards, inspired in the traditional housing from La Mancha. Above the base, looking to the road, as a media screen, the black box emerges as a symbolic and representative element, wrapped in mobile sheets of blue polycarbonate.

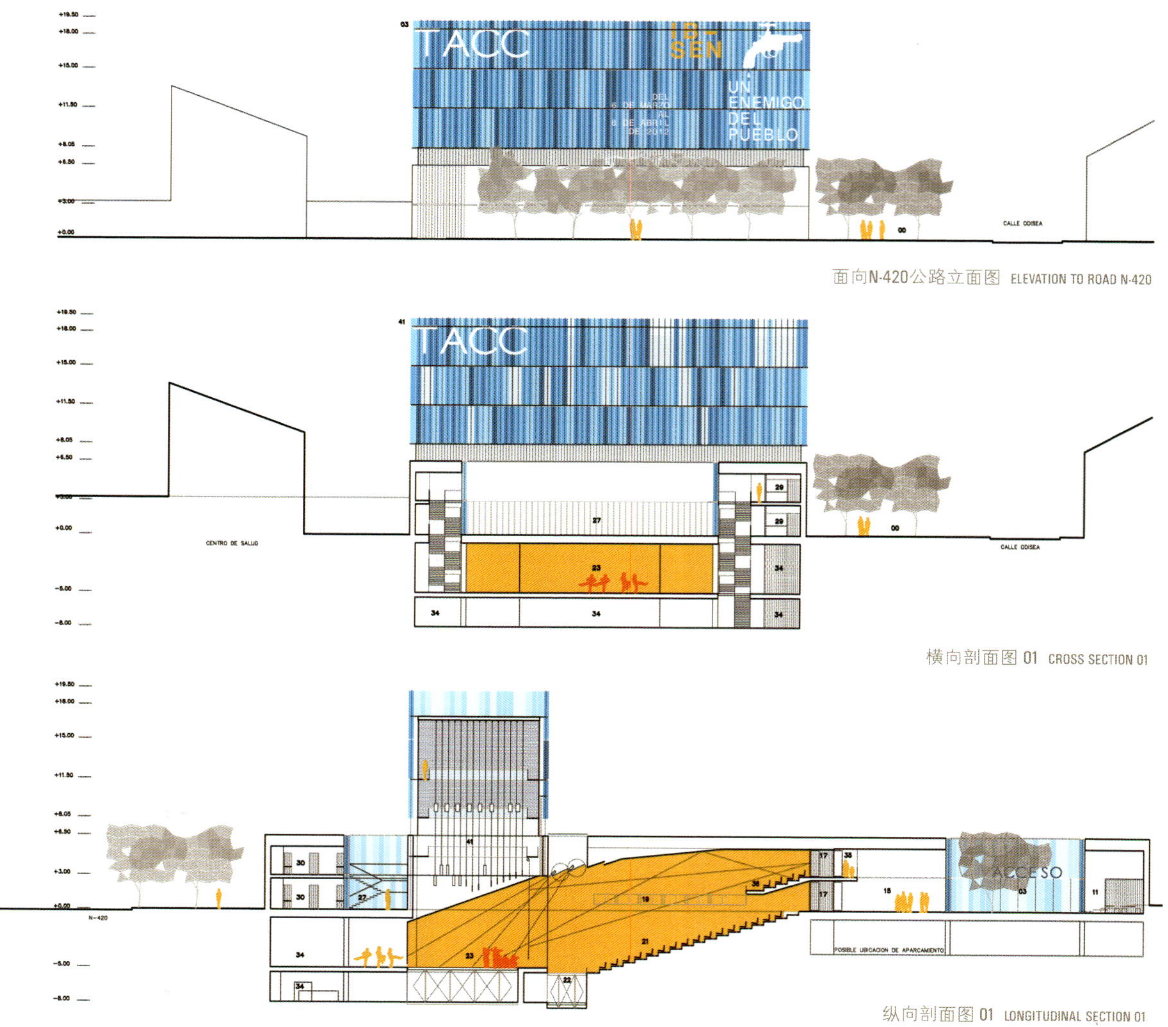

面向N-420公路立面图 ELEVATION TO ROAD N-420

横向剖面图 01 CROSS SECTION 01

纵向剖面图 01 LONGITUDINAL SECTION 01

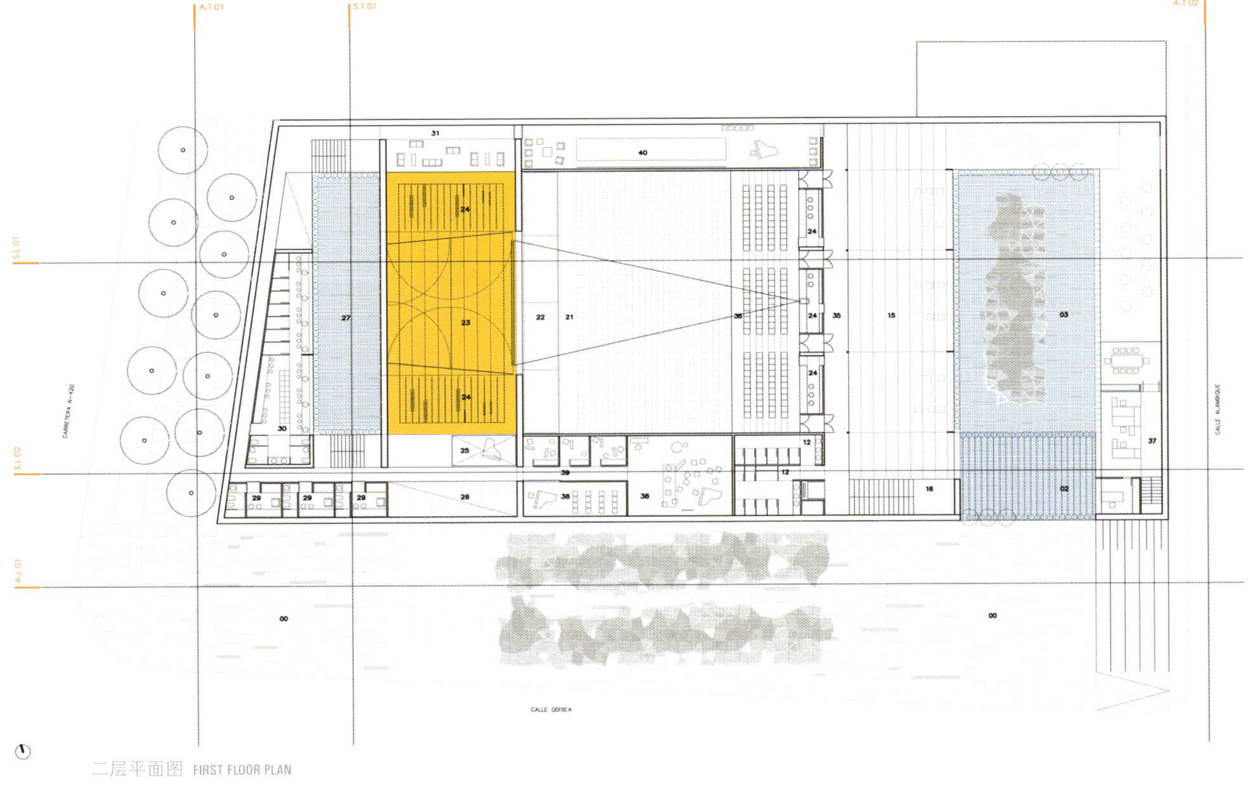

二层平面图 FIRST FLOOR PLAN

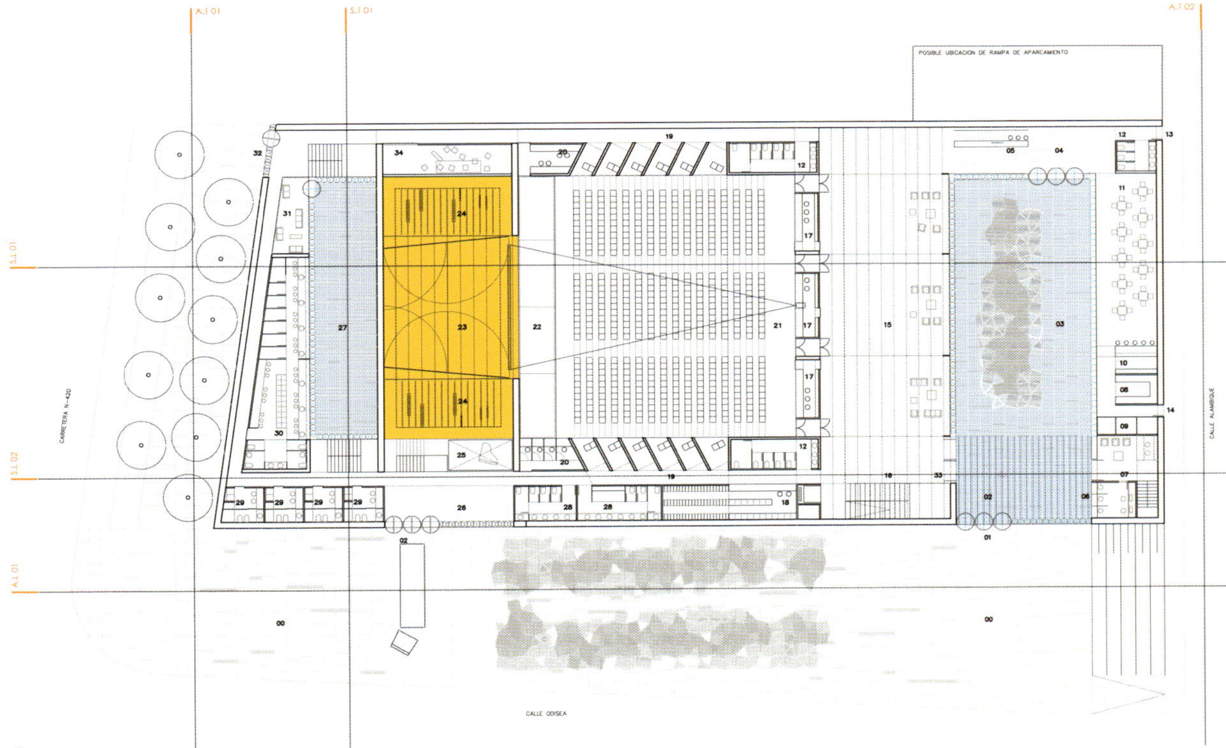

入口层平面图 ACCESS FLOOR PLAN

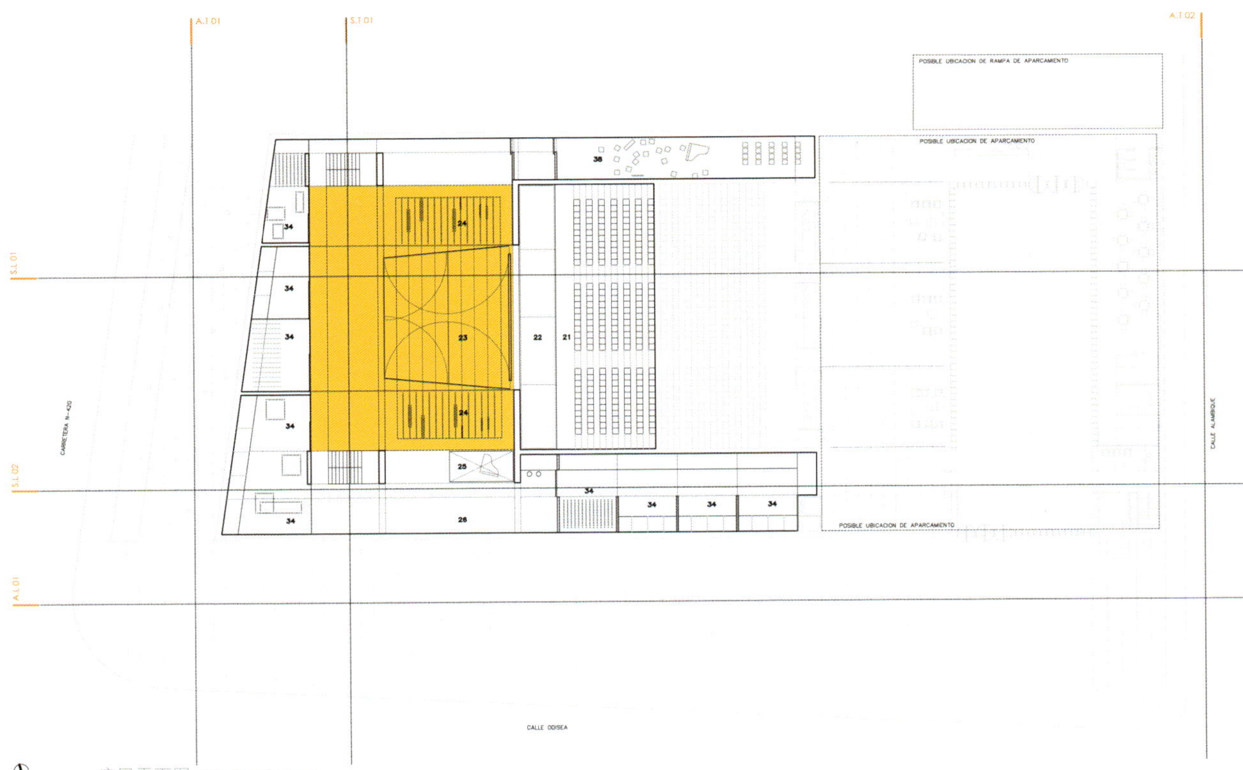

底层平面图 GROUND FLOOR PLAN

填充与空地 SOLIDS AND VOIDS

填充地带 SOLID SPACES

空地 EMPTY SPACES

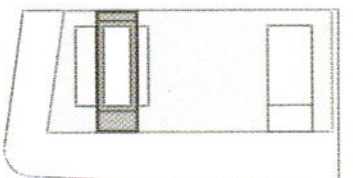

+2 演艺厅 BLACK BOX

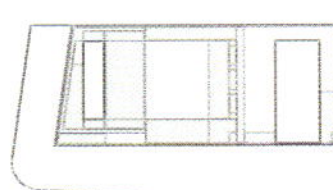

+1

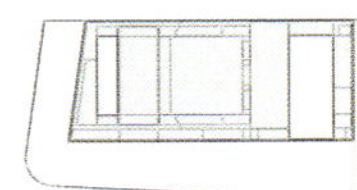

+0

独立及公共空间 PUBLIC AND PRIVATE SPACES

公共空间 PUBLIC SPACES

独立空间 PRIVATE SPACES

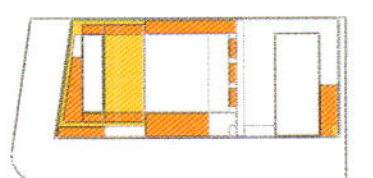

+2 演艺厅 BLACK BOX

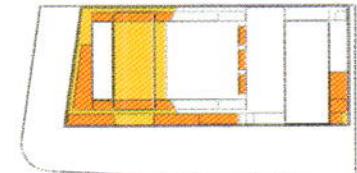

+1

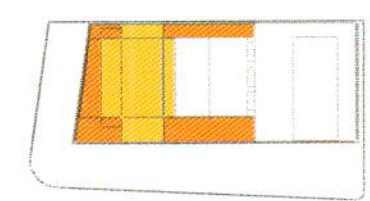

+0

通道示意图 DIAGRAM OF ROUTES

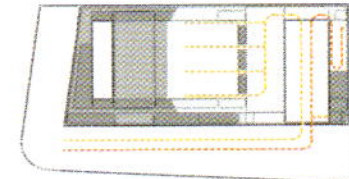

公共通道・礼堂的开放及关闭 PUBLIC ROUTE · OPEN + CLOSED AUDITORIUM

公共空间 PUBLIC SPACES

独立空间 PRIVATE SPACES

开放的公共通道 PUBLIC ROUTE. OPEN

关闭的公共通道 PUBLIC ROUTE. CLOSED

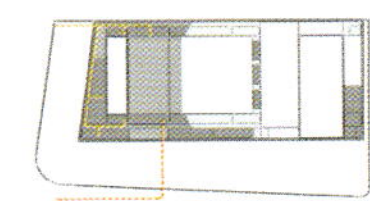

表演者通道 + 装卸区 ARTISTS ROUTE + LOAD DOWNLOAD

公共空间 PUBLIC SPACES

独立空间 PRIVATE SPACES

表演者通道 ARTISTS ROUTE

装卸通道 LOAD DOWNLOAD ROUTE

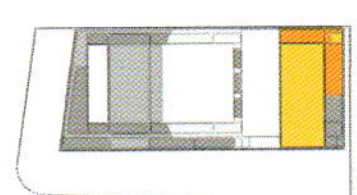

表演时段外的开放空间 OPEN AREAS OUT OF SHOW TIMETABLE

公共空间 PUBLIC SPACES

独立空间 PRIVATE SPACES

表演时段外公共空间 PUBLIC SPACES OPEN OUT OF TIMETABLE

克里普塔纳乡村剧院礼堂 · 雷阿尔城

Campo de Criptana Theatre-Auditorium · Spain

参标 · Proposal

entreacto (竞标代码)

Eva Iszoro (建筑师)

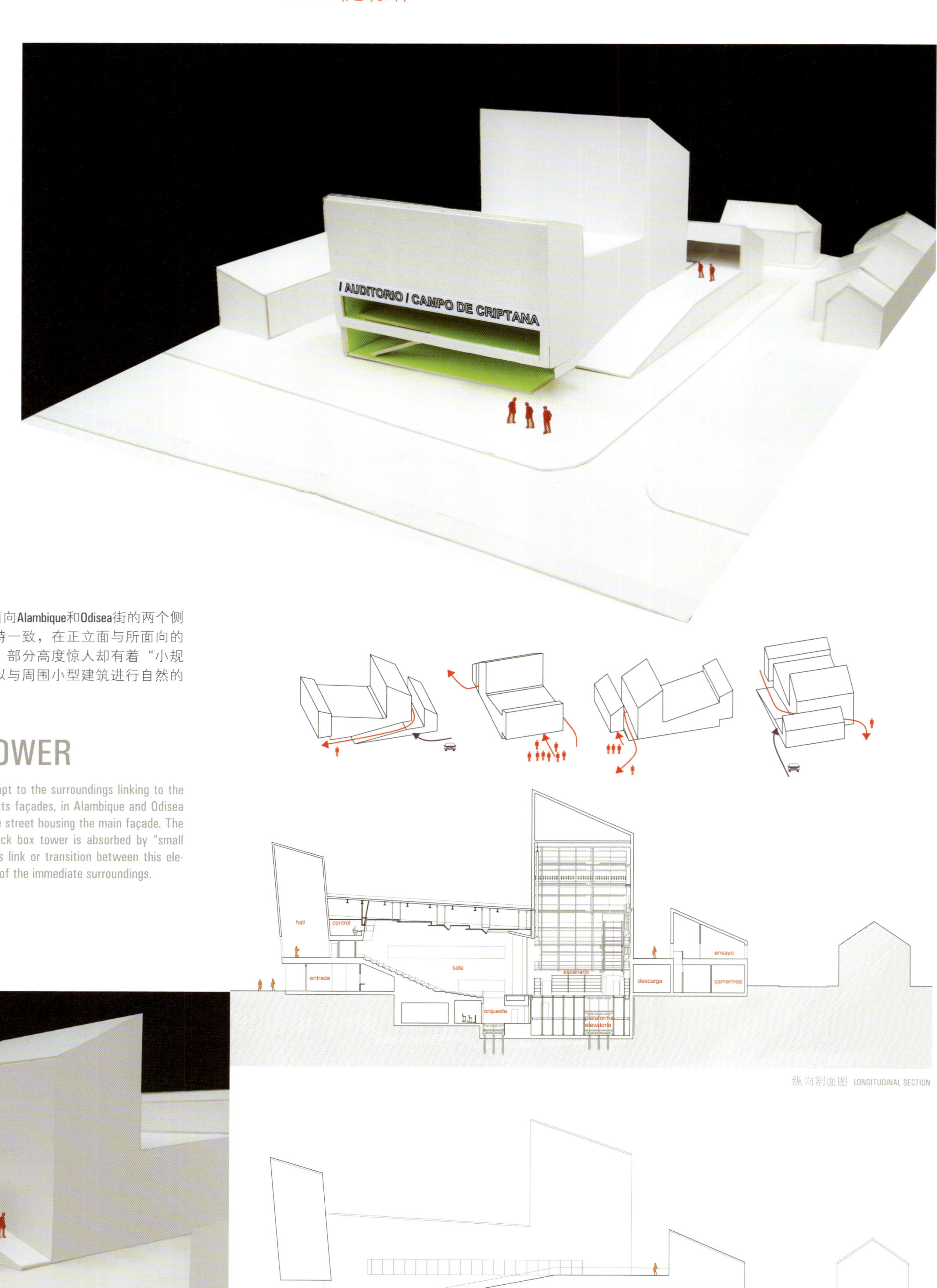

纵向剖面图 LONGITUDINAL SECTION

ALAMBIQUE大街立面图 ELEVATION TO ALAMBIQUE STREET

隐楼

表面上，该建筑面向Alambique和Odisea街的两个侧立面要与周围建筑保持一致，在正立面与所面向的街道分离。“黑盒子”部分高度惊人却有着“小规模”体积，这样就可以与周围小型建筑进行自然的衔接或过渡。

HIDDEN TOWER

The building pretends to adapt to the surroundings linking to the existing typology in two of its façades, in Alambique and Odisea streets, and dissociate in the street housing the main façade. The impressive height of the black box tower is absorbed by "small scale" volumes which act as link or transition between this element and the little buildings of the immediate surroundings.

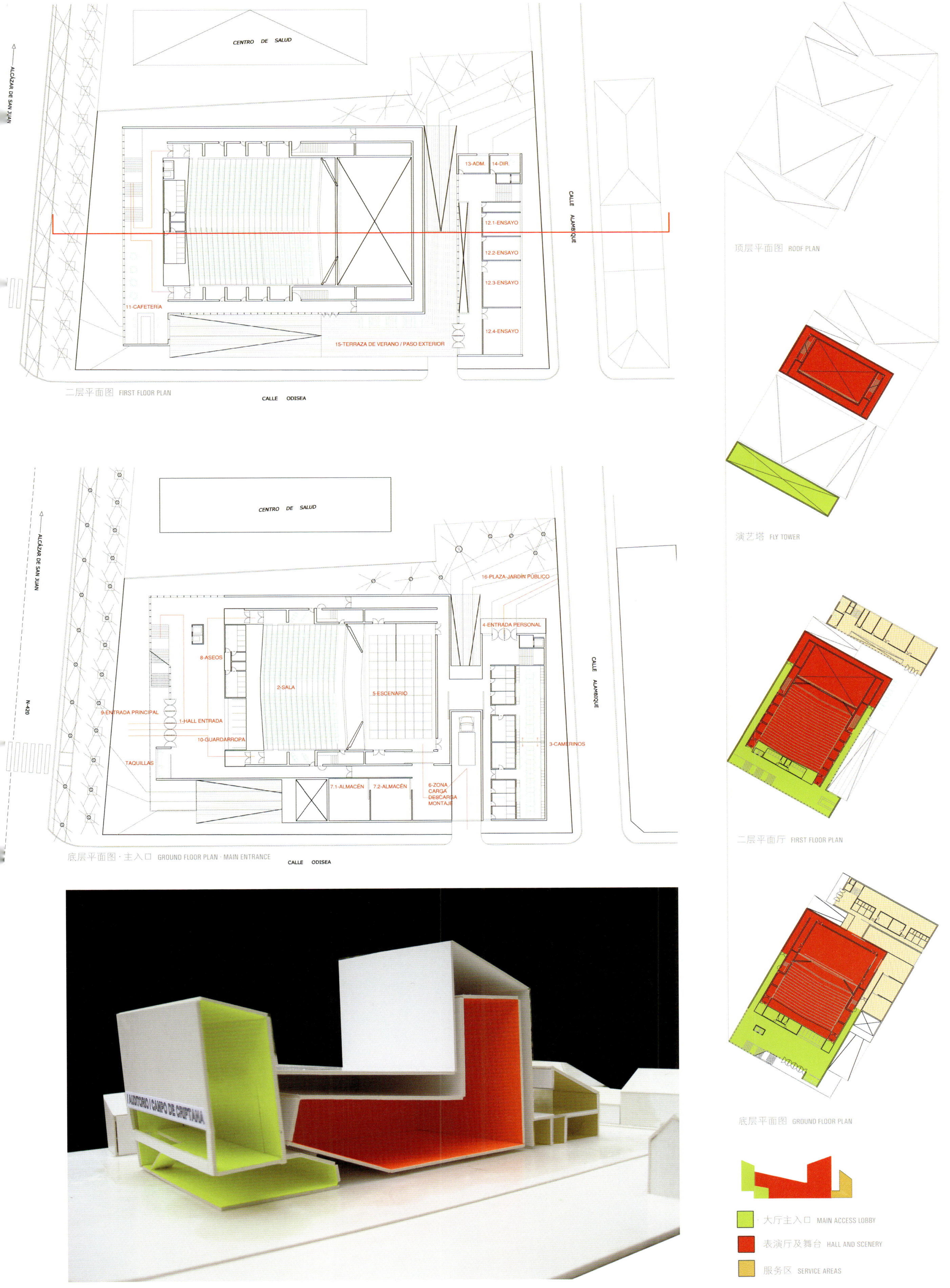

二层平面图 FIRST FLOOR PLAN

底层平面图 · 主入口 GROUND FLOOR PLAN · MAIN ENTRANCE

顶层平面图 ROOF PLAN

演艺塔 FLY TOWER

二层平面厅 FIRST FLOOR PLAN

底层平面图 GROUND FLOOR PLAN

大厅主入口 MAIN ACCESS LOBBY

表演厅及舞台 HALL AND SCENERY

服务区 SERVICE AREAS

原岛屿体育场都市花园 · 大加那利群岛拉斯帕尔玛斯

Urban Park of the Insular Stadium · Canary Islands, Spain

竞标 · competition
原岛屿体育场都市花园
Urban Park in the existing Insular Stadium

竞标类型 · competition type
一轮制概念公开竞标
open, one-stage design competition

项目地点 · site area
大加那利群岛拉斯帕尔玛斯 · 西班牙 Las Palmas de Gran Canaria · Spain

主办方 · promoter
大加那利群岛市政府及大加那利群岛拉斯帕尔玛斯市政府
Cabildo de Gran Canaria y Ayuntamiento de Las Palmas de Gran Canaria

日程安排 · schedule
招标 · Announcement 09.2008
评审结果 · Jury´s results 01.2009

评审团 · jury

主席 presidente:
Emilio Mayoral Fernández
委员 vocales:
Demetrio Suárez Díaz
Antonia García Carló
Juan Ramírez Guedes
Héctor García Sánchez
Pilar Banqueri Sánchez
Manuel Martín Hernández

获奖者 · awards

中标 · winner
Casariego / Guerra arquitectos (建筑师事务所)
Elsa Guerra Jiménez · Joaquín Casariego Ramírez

合作 (c) Noemi Tejera · Rubén Pérez · Alejandro Morán · Silvia Guajardo · Marta Cuyás · Carmelo García · Juan Manuel Rivero · Denébola Acosta

第一提名奖 · first accesit
Jetworks arquitectos (建筑师事务所)
Manuel J. Feo Ojeda · Claudia Collmar · Antonio Rocha Quintero

合作 (c) Jose Gregorio de Haro · David Rocha · Oscar de Castro · Alina Valcarce · Javier Barceló · Catherina Castro · Francisco Javier Méndez

第二提名奖 · second accesit
Oscar Rebollo Curbelo (建筑师)
Eva Padrón Sánchez (建筑师)
Cinta Rodríguez Cabrera (建筑师)

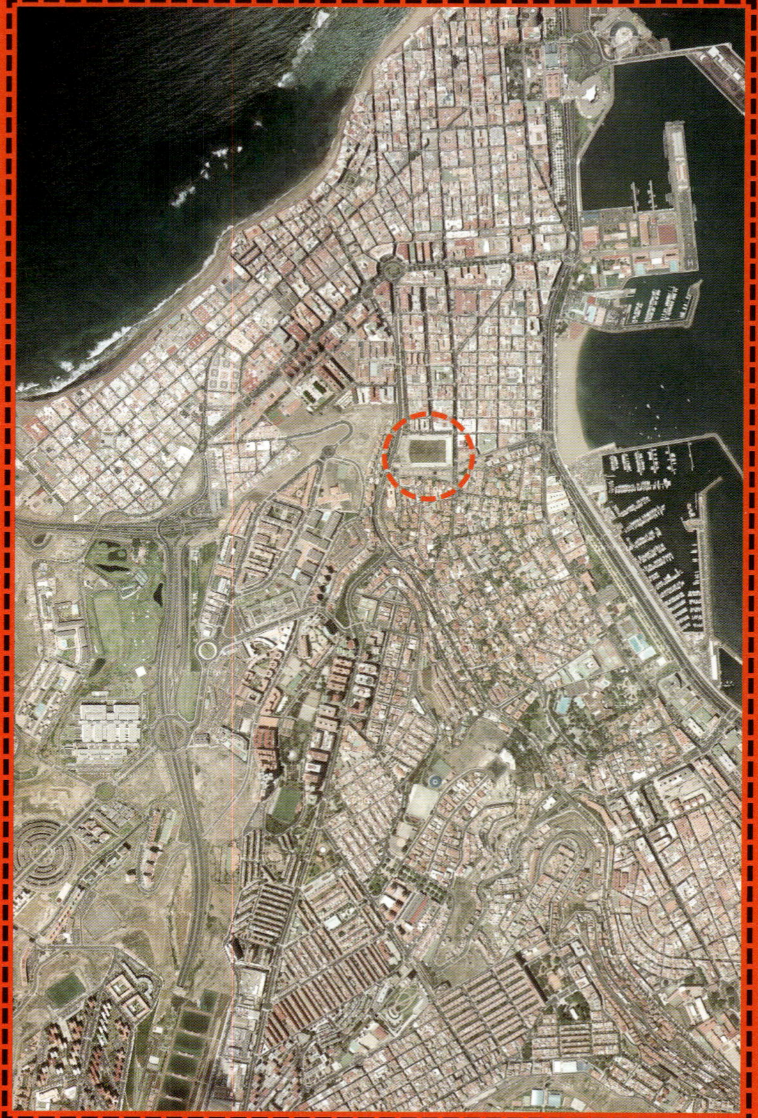

出于竞赛的需要，**我们一方面需要凸显北面和东面的建筑价值**（面向Pio XII），另外一方面修整并恢复使用具有重大历史意义的曲形看台。同时，竞赛旨在在建筑物南面设计一个广场，该广场能将整个建筑空间和一个大的绿色空间整合在一起。

The substantial support of the architectural value of the north façade and east façade (facing Pio XII), and the recovery and use of the trace of the "curve stand" due to its historical significance were among the requirements of the competition. At the same time, the competition looked for the design of the square south to the building which could integrate the whole space and a big green space.

原岛屿体育场都市花园 · 拉斯帕尔玛斯

Urban Park in the existing Insular Stadium · Spain

一等奖 · First Prize

030908ei (竞标代码)

Casariego / Guerra arquitectos (建筑师事务所)

Elsa Guerra Jiménez · Joaquín Casariego Ramírez (建筑师)

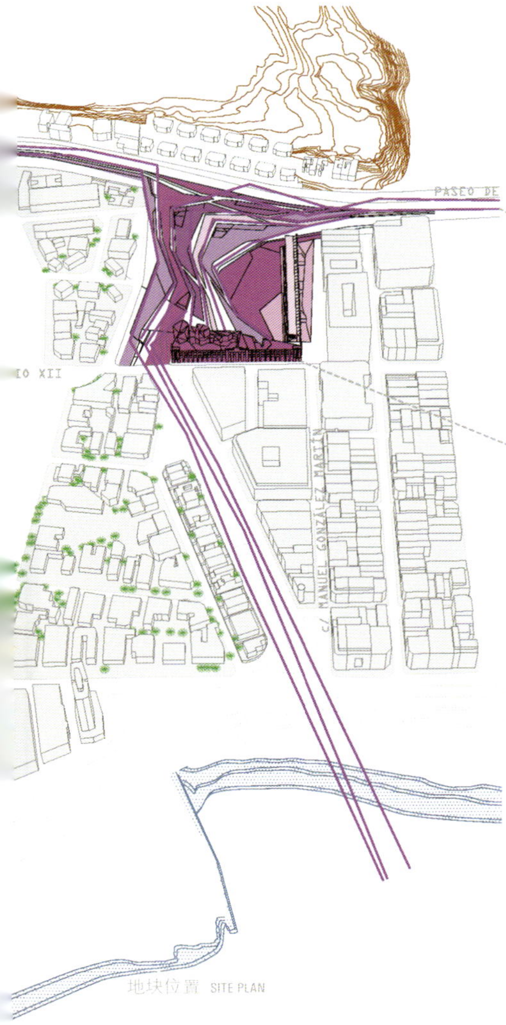

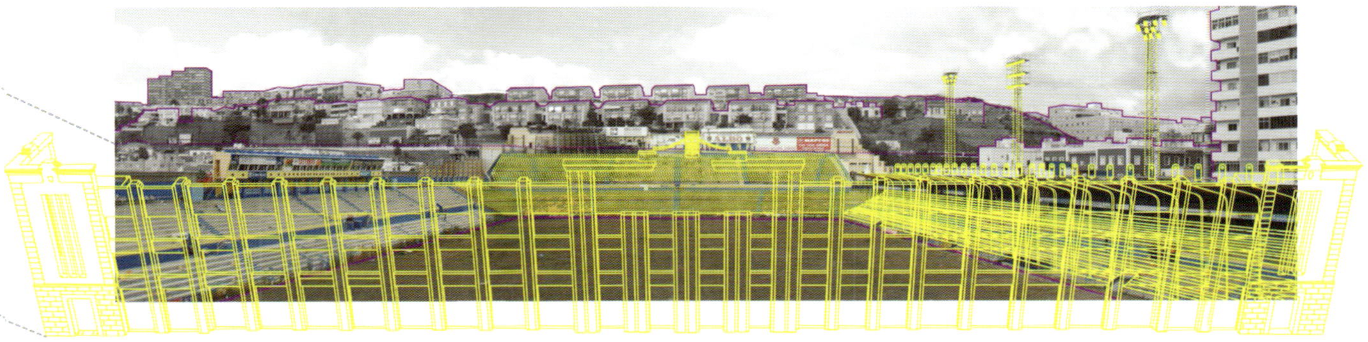

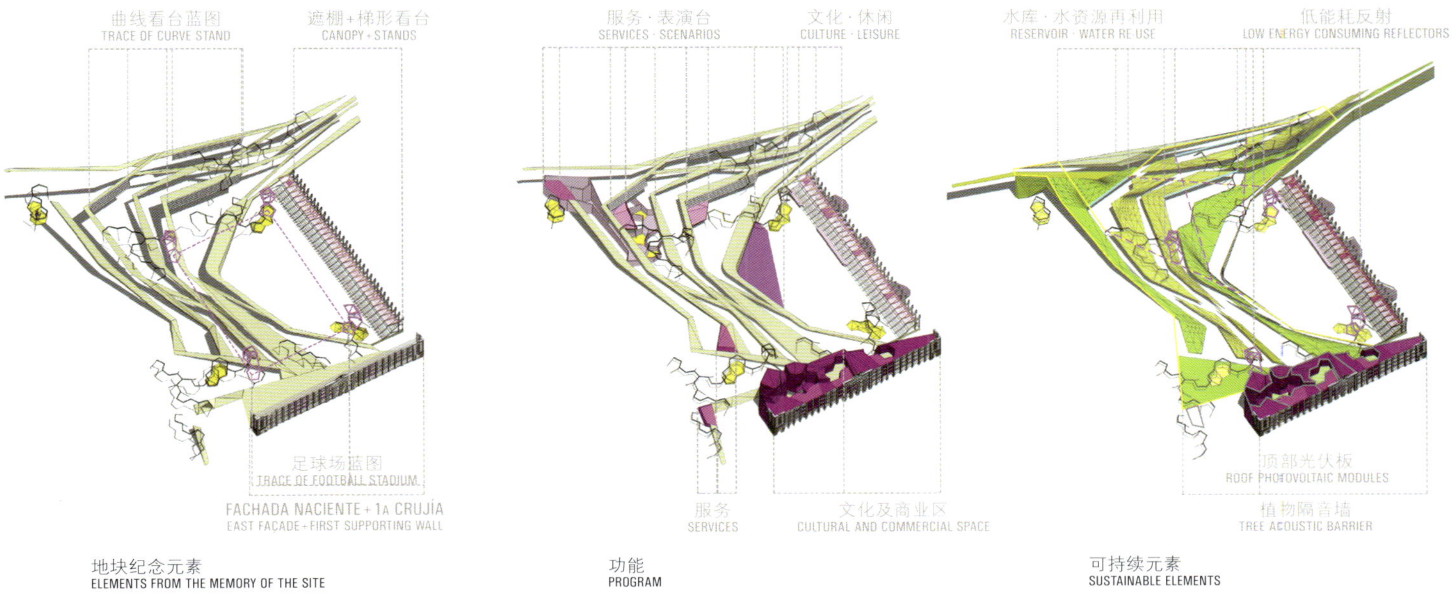

地块位置 SITE PLAN

曲线看台蓝图 TRACE OF CURVE STAND

遮棚+梯形看台 CANOPY+STANDS

足球场蓝图 TRACE OF FOOTBALL STADIUM

FACHADA NACIENTE+1A CRUJÍA EAST FACADE+FIRST SUPPORTING WALL

地块纪念元素 ELEMENTS FROM THE MEMORY OF THE SITE

服务 · 表演台 SERVICES · SCENARIOS

文化 · 休闲 CULTURE · LEISURE

服务 SERVICES

文化及商业区 CULTURAL AND COMMERCIAL SPACE

功能 PROGRAM

水库 · 水资源再利用 RESERVOIR · WATER RE-USE

低能耗反射 LOW ENERGY CONSUMING REFLECTORS

顶部光伏板 ROOF PHOTOVOLTAIC MODULES

植物隔音墙 TREE ACOUSTIC BARRIER

可持续元素 SUSTAINABLE ELEMENTS

足球忆事

FOOTBALL REMINISCENCE

覆盖整个公园的照明线路系统是基于足球的几何外形设计而成的，该系统经由一个六边形将一个个五边形连接起来，并同时可在其中构建娱乐项目及为攀爬植物的自由生长提供环境。体育场目前仍保留着四个"角"，经光的折叠而显现出来。

The system of lighting Lines which cover the park is based on the geometry of the football, which connects pentagons from a hexagon, and at the same time shapes entertainment programs and allows the growth of climbing plants. The stadium is present, preserving the four "corners", pointed out by the fold of the light.

总平面图 OVERALL PLAN

空余空间 FREE SPACES

A. 绿地 GREEN AREAS

1 棕榈树沿线斜坡 SLOPES WITH PALM TREE LINES
2 爬行植物沿线斜坡 SLOPES WITH CLIMBING PLANTS
3 植物沿线斜坡 SLOPES WITH TREES
4 软路面走道 SOFT WALKWAYS
5 植物沿线过道 WALKWAYS WITH TREES
6 花园区域 GREEN AREAS
7 看台 STANDS
8 次要表演台 SECONDARY SCENARIOS
9 停留区 LONG STAY ZONES
10 儿童游戏区 ZONES FOR CHILDRENS GAMES

B. 路面区域 PAVED AREAS

11 主要表演台 MAIN SCENARIO
12 广场 PLAZA
13 看台 STANDS
14 观景台 VIEW-POINT
15 地下入口 UNDERGROUND ACCESS
16 六配五边形小广场 HEXA-PENTAGONAL LITTLE SQUARES
17 人行过道 PEDESTRIAN WALKWAYS
连续路面 CONTINUOUS PAVEMENT
岩石路面 PETROUS PAVEMENT
鹅卵石路面 PAVING STONE PAVEMENT

C. 沿线照明 · 游戏 LIGHT LINES · GAMES

D. 角落 CORNER

E. 运动区 SPORTS AREAS

18 壁球场 SQUASH COURTS
19 多用途场地 MULTIPLE COURTS
20 自行车租借 BICYCLE LOANING

设备 EQUIPMENTS

F. "UD中心" "UD CENTRE"

服务设备区 SERVICE BUILDING
商业区 COMMERCIAL AREA
文化区 CULTURAL AREA

G. "论坛广场" "TRIBUNA SQUARE"

亭子 KIOSKS
服务 · 活动 SERVICES · EVENTS
社会文化 SOCIO-CULTURAL

H. 服务 SERVICES

21 体育 SPORTS
22 维护 MAINTENANCE

I. 地下停车场 UNDERGROUND PARKING

23 汽车入口 CAR ACCESS

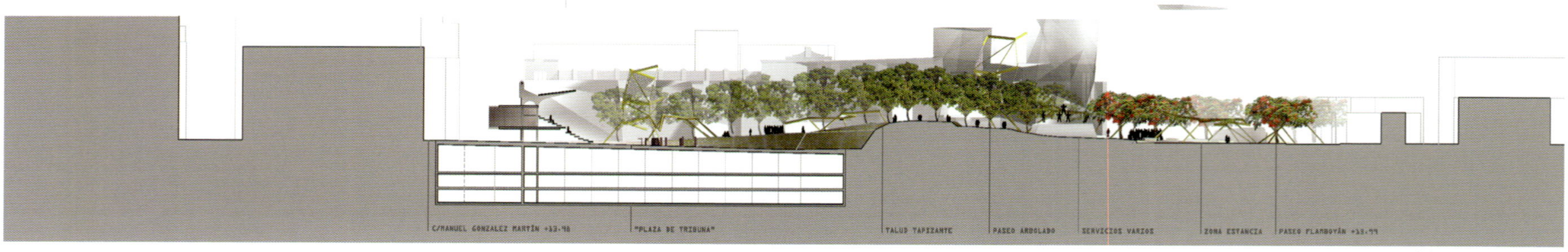

剖面图 AA' SECTION AA'

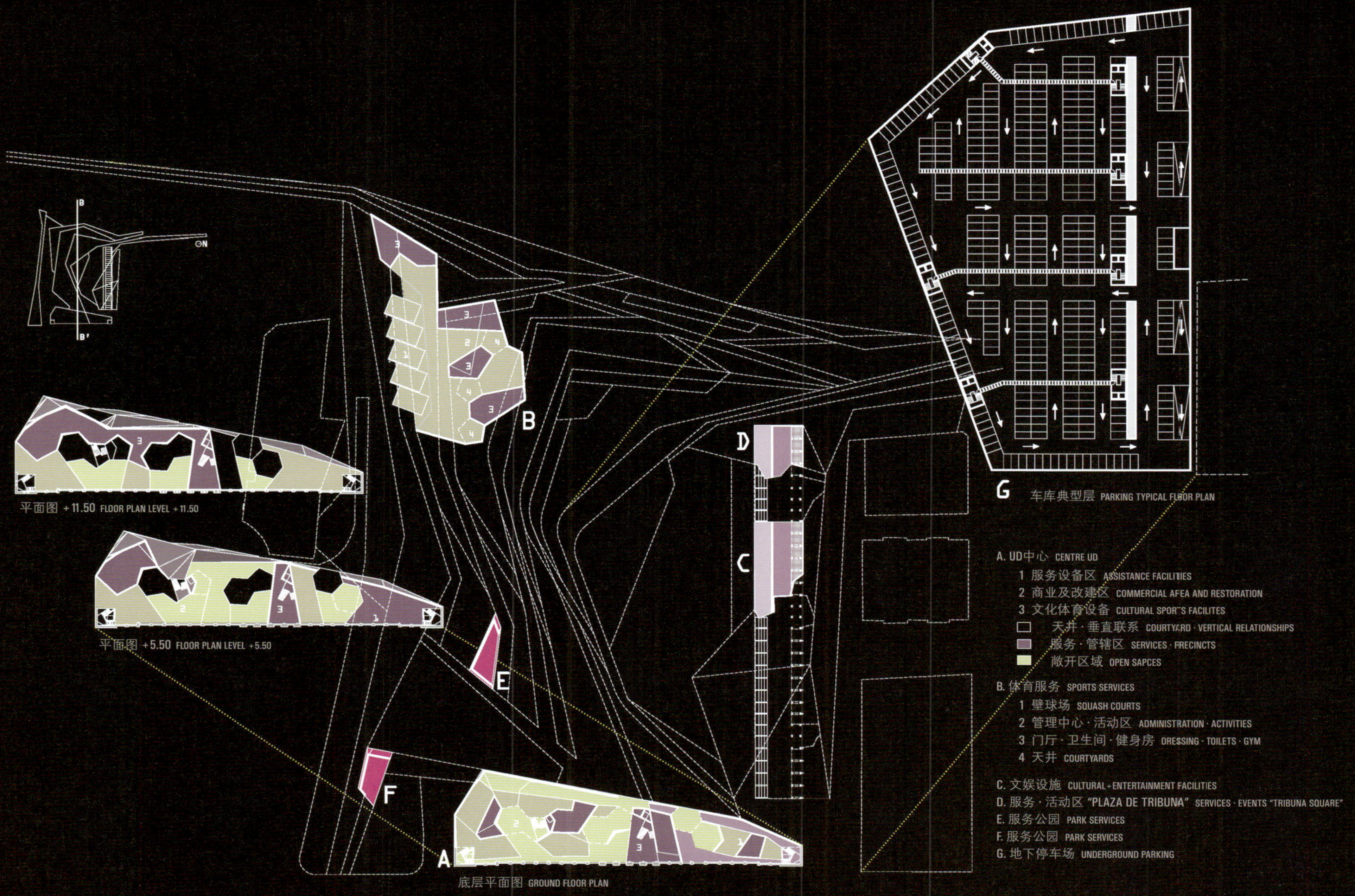

平面图 +11.50 FLOOR PLAN LEVEL +11.50

平面图 +5.50 FLOOR PLAN LEVEL +5.50

底层平面图 GROUND FLOOR PLAN

G 车库典型层 PARKING TYPICAL FLOOR PLAN

A. UD中心 CENTRE UD
- 1 服务设备区 ASSISTANCE FACILITIES
- 2 商业及改建区 COMMERCIAL AFEA AND RESTORATION
- 3 文化体育设备 CULTURAL SPOR'S FACILITES
- 天井 · 垂直联系 COURTYARD · VERTICAL RELATIONSHIPS
- 服务 · 管辖区 SERVICES · PRECINCTS
- 敞开区域 OPEN SAPCES

B. 体育服务 SPORTS SERVICES
- 1 壁球场 SQUASH COURTS
- 2 管理中心 · 活动区 ADMINISTRATION · ACTIVITIES
- 3 门厅 · 卫生间 · 健身房 DRESSING · TOILETS · GYM
- 4 天井 COURTYARDS

C. 文娱设施 CULTURAL + ENTERTAINMENT FACILITIES
D. 服务 · 活动区 "PLAZA DE TRIBUNA" SERVICES · EVENTS "TRIBUNA SQUARE"
E. 服务公园 PARK SERVICES
F. 服务公园 PARK SERVICES
G. 地下停车场 UNDERGROUND PARKING

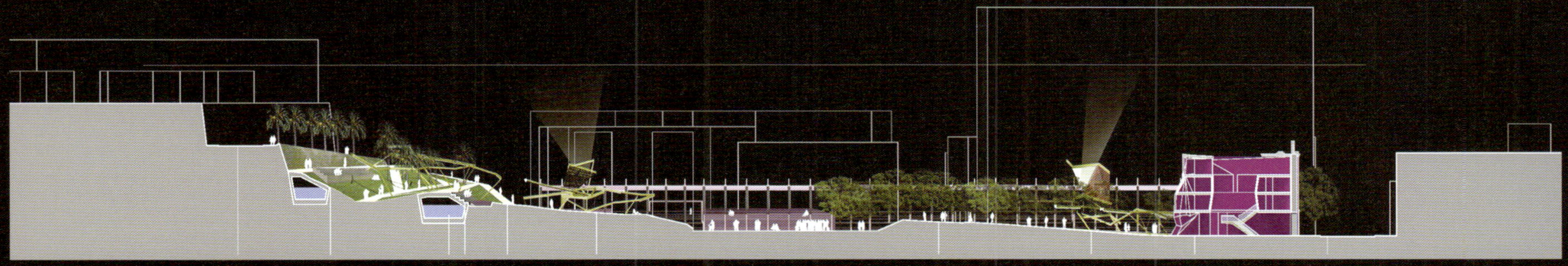

剖面图 BB′ SECTION BB′

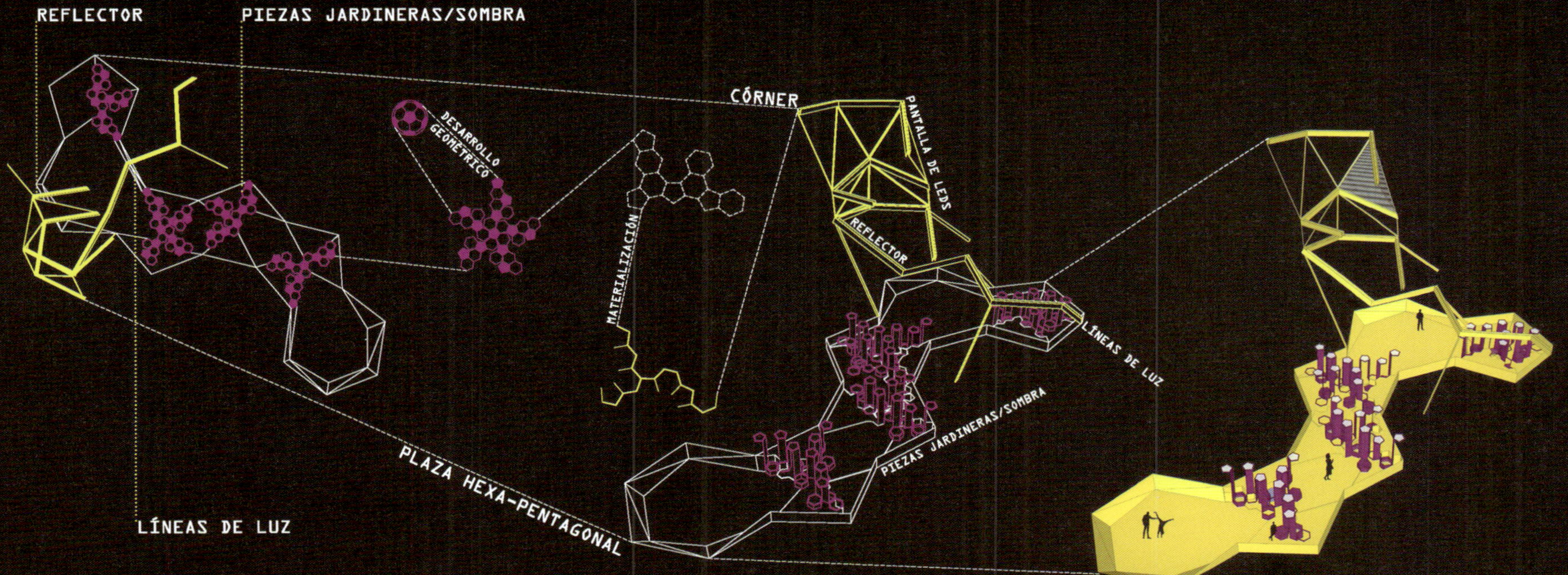

原岛屿体育场都市花园 · 拉斯帕尔玛斯

Urban Park in the existing Insular Stadium · Spain

第一提名奖 · First Accesit

fr381362 (竞标代码)

Jetworks arquitectos (建筑师事务所)

重塑传统

体育场视野较宽，能见度较高，人们在这里进行体育运动并释放更多的激情，比如踢足球。将岛状体育场改建成城市公园，在改造过程中，记忆被施工材料一点点吞噬。我们并不想把顶棚单一地建成考古构架，它必将成为线性屋顶住房文化项目的一部分，拥有图书馆、教室、书店和展览馆。

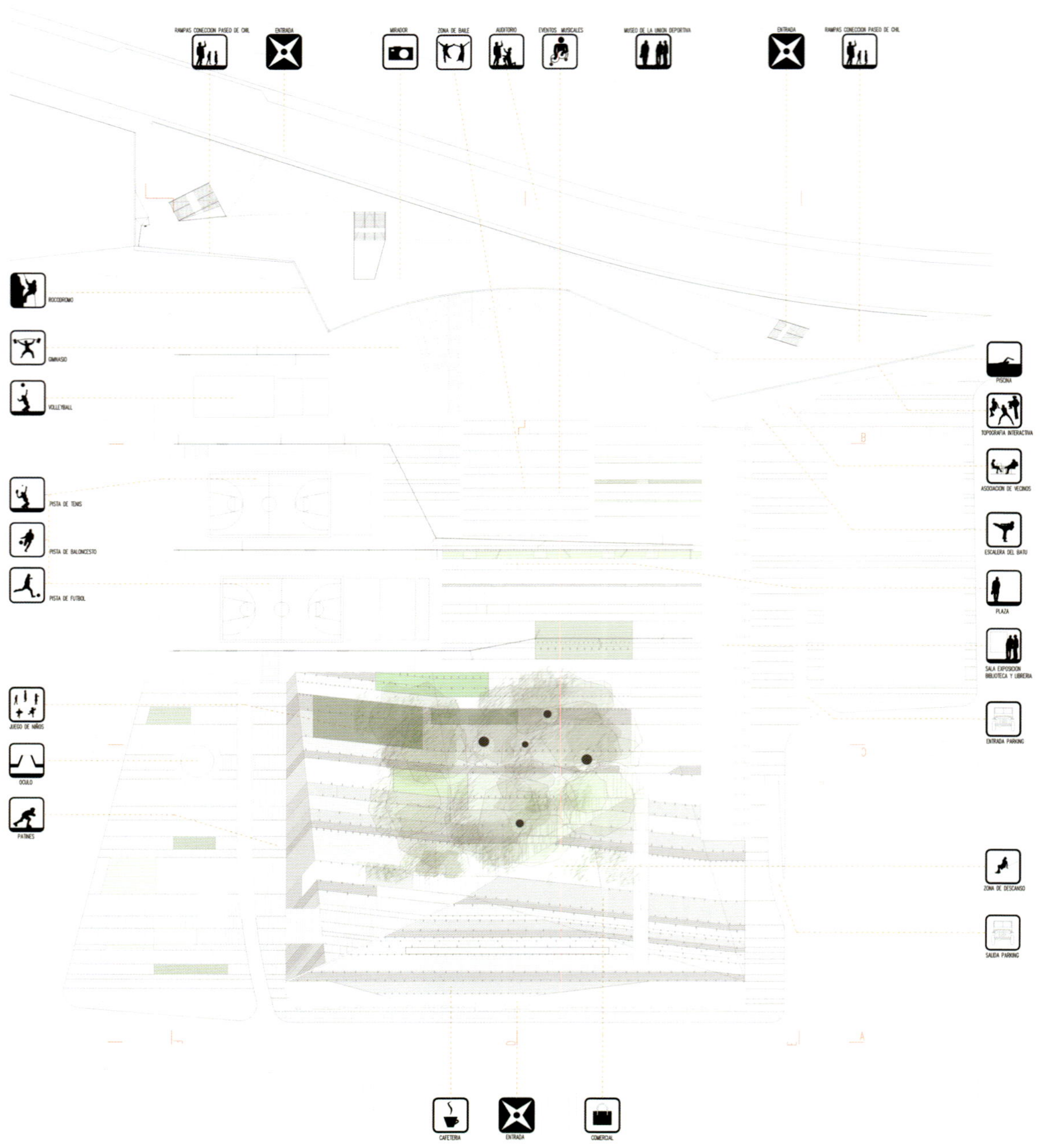

THE REINVENTION OF THE TRADITION

A stadium is a machine to create visibility, an artefact that intensifies and shows the emotions which are generated practising a sport like football. The conversion of the insular stadium into an urban park promotes to work in a place where the memory turns into construction material. We do not want to have the canopy acting only as archaeological skeleton; it must be part of the linear roof housing cultural programs: library, classrooms, bookshop and exhibition.

顶层平面图 ROOF PLAN

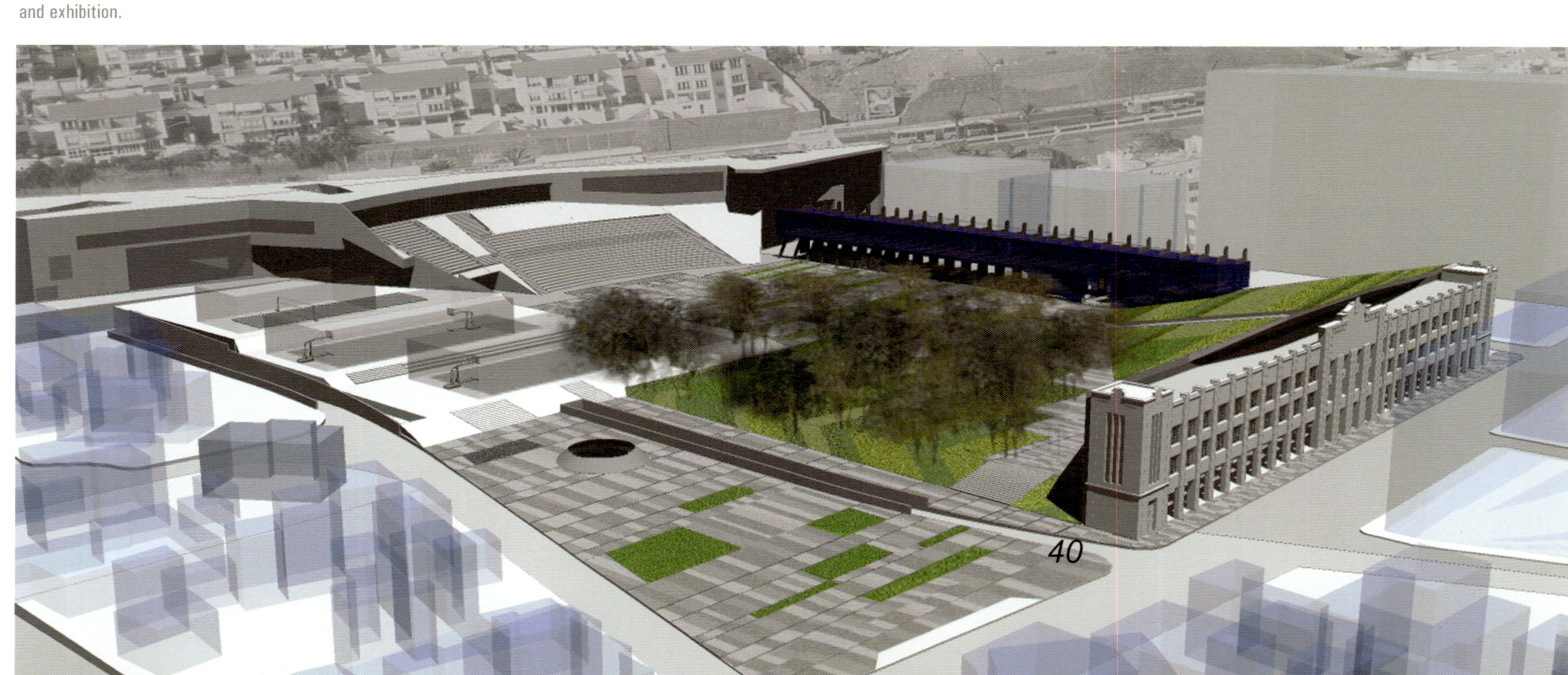

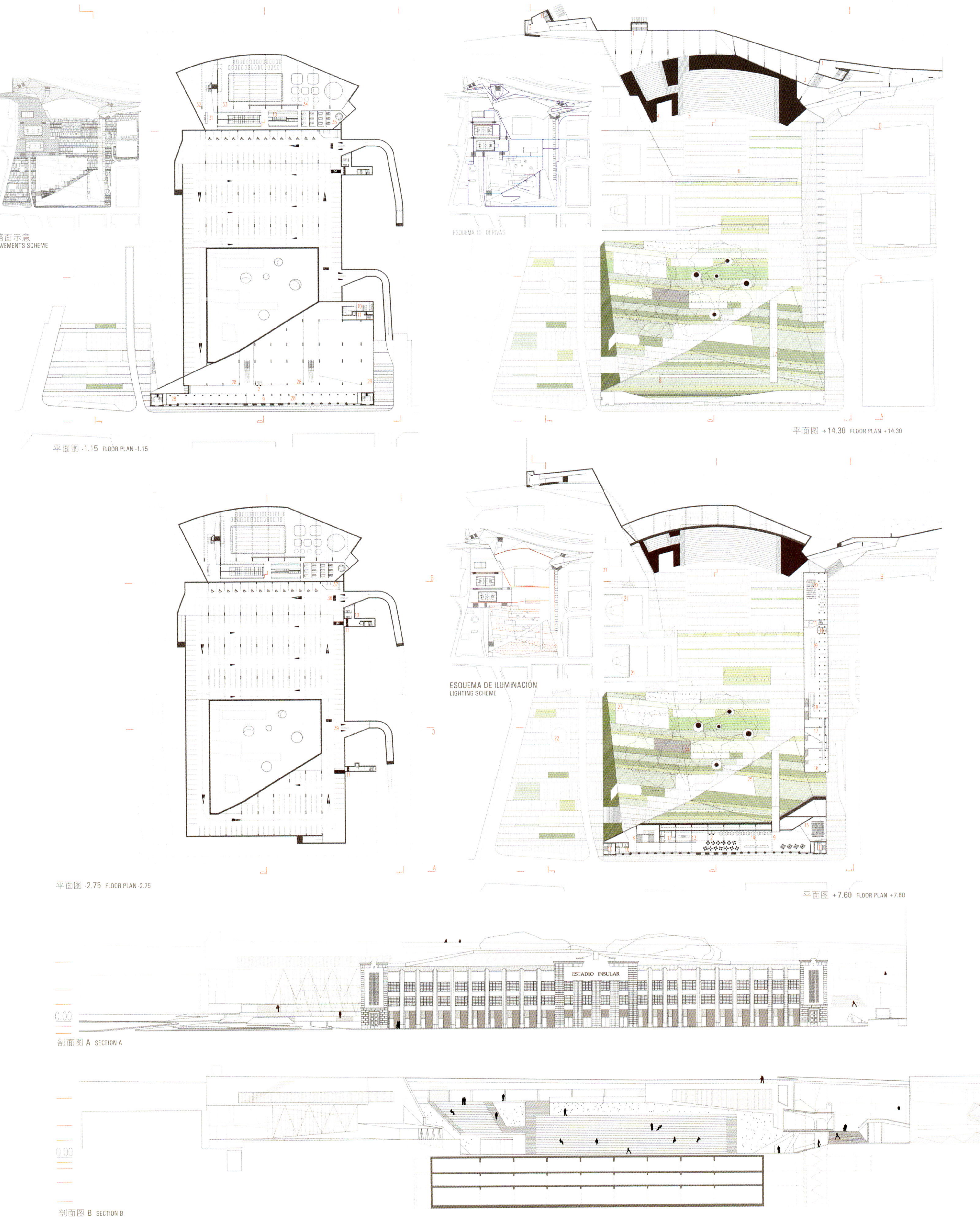

平面图 -1.15 FLOOR PLAN -1.15

平面图 +14.30 FLOOR PLAN +14.30

平面图 -2.75 FLOOR PLAN -2.75

平面图 +7.60 FLOOR PLAN +7.60

剖面图 A SECTION A

剖面图 B SECTION B

原岛屿体育场都市花园 · 拉斯帕尔玛斯

Urban Park in the existing Insular Stadium · Spain

第二提名奖 · Second Accesit

2812070e (竞标代码)

Óscar Rebollo Curbelo · Eva Padrón Sánchez · Cinta Rodríguez Cabrera (建筑师事务所)

主要连接及路线 MAIN CONNECTIONS AND ROUTES

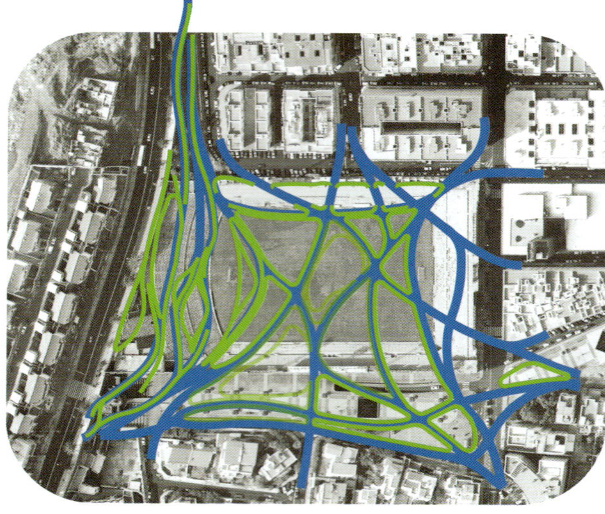

相连空间 RELATIONSHIP SPACES

人流曲线 FLUX DIAGRAMS

新地形

地形已逐渐成为居民新型生活的本质体现和基本物质要求。我们建议在岛状体育场的城市公园中引进流体空间，将其与周边所有邻近地区都连接起来，从而创建新的地形，并借机使修建大型庭院、还原体育场的历史、修补现有建筑、建立新项目以及增加新建筑得以同时完成，以满足居民需求。

A NEW TOPOGRAPHY

The topography turns into an element designed as essence, basic substance of new relationships in the citizens´ lives. We propose to introduce a fluid space in the urban park of the insular stadium with connections between all surrounding neighbourhoods creating a new topography. It is an opportunity to recover the history of the stadium with the recovery of the existing building with new programs related to the needs of citizens and with the addition of a new building, connected with big courtyards.

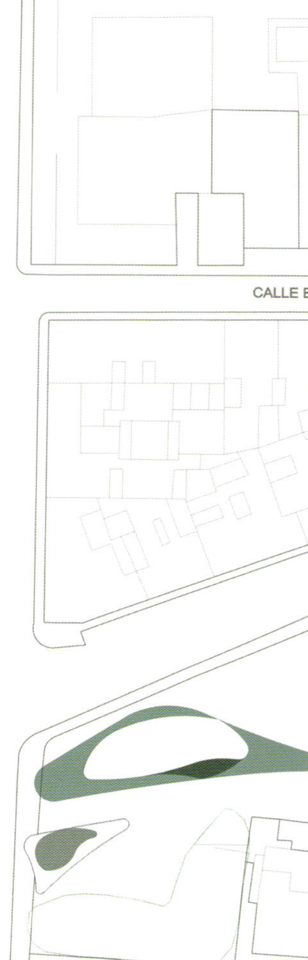

平面图 FLOOR PLAN

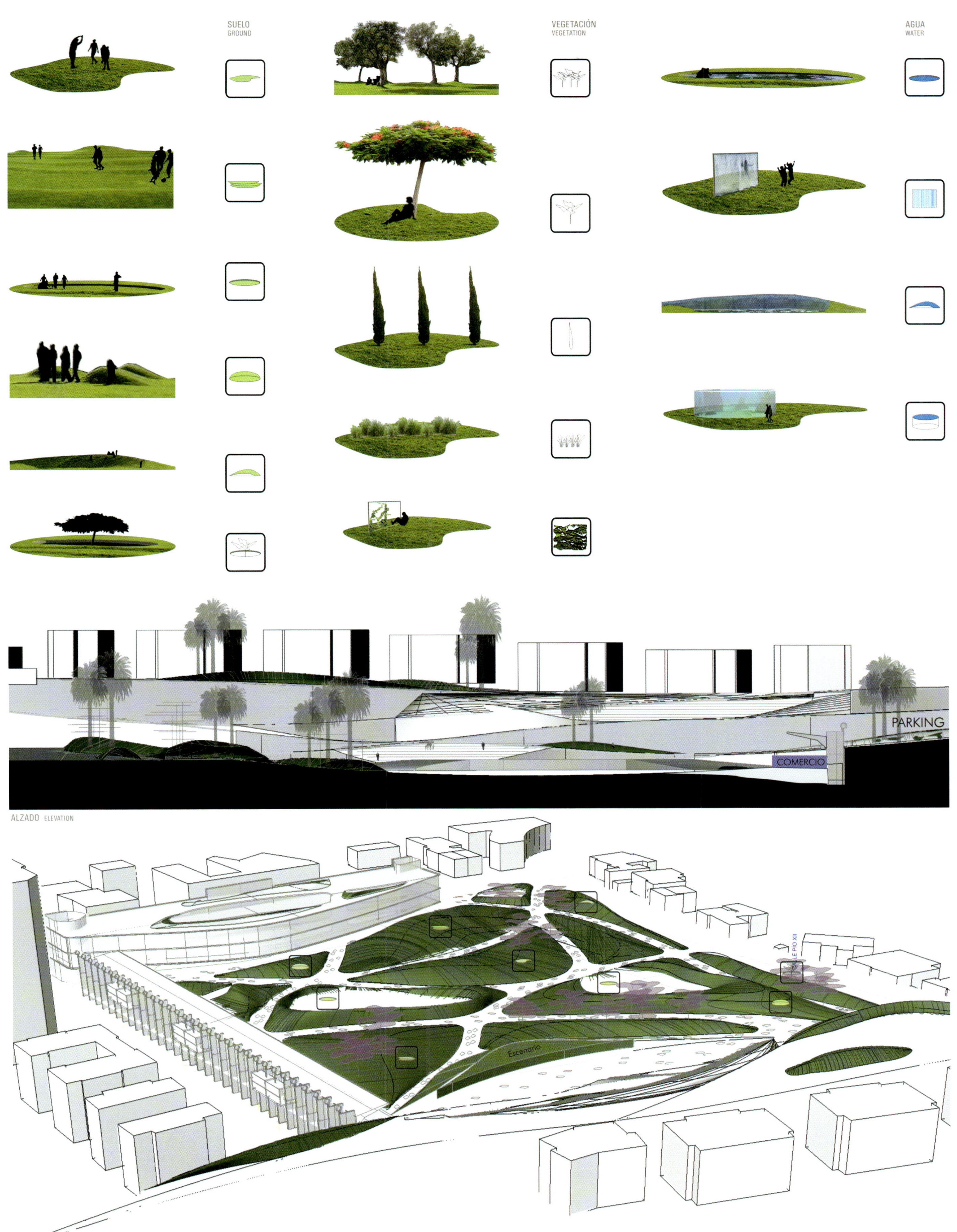

SUELO
GROUND
VEGETACIÓN
VEGETATION
AGUA
WATER
PARKING
COMERCIO
ALZADO ELEVATION
Escenario

苏黎世国际机场航站大楼 "The Circle"

"The Circle" at Zurich Airport · Switzerland

竞标 · competition
苏黎世国际机场航站大楼
"The Circle" at Zurich Airport

竞标类型 · competition type
三轮制概念公开竞标
open, three-stage design competition

项目地点 · site area
苏黎世 · 瑞士 Zürich · Switzerland

主办方 · promoter
Unique Flughafen zürich AG 公司

日程安排 · schedule

招标 · Announcement	04.2009
评审结果 · Jury´s results	01.2010

评审团 · jury

Andreas Schmid	Stefan Bitterli
Thomas E. Kern	Regine Leibinger
Stephan Widrig	Jürgen Mayer H.
Peter Eriksson	Werner Sobek
Arthur Tobler	Wiel Arets
René Huber	Kees Christiaanse
Carl Fingerhuth	Birgit Werner

获奖者 · awards

一等奖 · first prize
Riken Yamamoto & Field Shop (建筑师事务所)
Riken Yamamoto & Beda Faessler Architect · CH-Architekten
结构 engineering: Arup Japan · Basler & Hofmann

二等奖 · second prize
Zaha Hadid Architects (建筑师事务所)
设计 design: Zaha Hadid and Patrik Schumacher
主案联合建筑师 project Architect · Design Associate: Manuela Gatto
技术联合 technical associate: Dillon Lin

三等奖 · third prize
Xaveer De Geyter Architecten · Bureau Dan Budik
(建筑师事务所)

交通工程 traffic engineer: Ernst Basler and Partner
结构工程 structural engineer: Ney & Partners
现场主管 construction management: B+P Baurealisation

四等奖 · fourth prize
Asymptote Architecture (建筑师事务所)

五等奖 · fifth prize
Dürig AG (建筑师事务所)
Jean-Pierre Dürig · Gian Paolo Ermolli · Stefania Koller · Philipp Schaefle · Matthias Winter · Claudia Lehmann · Joana Pereira Santos · Oliver Vogler (建筑师)

入围 · finalist
EM2N | Mathias Müller | Daniel Niggli
Architekten AG | ETH | SIA | BSA (建筑师事务所)
Daniel Niggli · Mathias Müller (建筑师)

团队 team: Duarte Brito · David Brodbeck · Takumi Iyoda · Fabian Hörmann · Mathias Kampmann Tim Klauser · Yoshi Nagamine · Yurika Orita · Indra Santosa · Martin Schriener · Jorrit Verduin
结构工程交通规划 structural engineer and traffic planning: weber + brönnimann ag
机械工程 mechanical engineer: Gruenberg + Partner AG
景观设计 landscape architects: Schweingruber Zulauf GmbH

入围 · finalist
Cruz y Ortiz arquitectos (建筑师事务所)
Antonio Cruz Villalón · Antonio Ortiz García (建筑师)

合作 (c) Jose Ortiz · Joaquín Hurtado · Curro Pérez · Mercedes Pérez · Rocío Peinado Juan Carlos Mulero · Marije Ter Steege · Daniel Rodriguez · Miguel Velasco infografías: Alejandro Álvarez
模型 maqueta: Benjamín Marchesoni · Marchesoni Modelli
联合建筑师 arquitectos colaboradores: Giraudi & Wettstein (Felix Wettstein · Sandra Giraudi)
合作 (c) Monica Delmenico · A. Bernini

苏黎世是瑞士最大的城市，也是该国商业、金融、高等教育和研究的中心地。苏黎世机场拥有2.4万名员工(2009)，每年的乘客量达2190万人。Unique (Flughafen Zürich AG)公司在苏黎世机场开发了一项新的项目，其中用于服务的高质量建筑的规划使用面积为20万平方米，包括到达航空终点站的步行距离。其被称为苏黎世国际机场航站大楼"The Circle"。此项目包括7个密切相连的模块：健康和美颜模块、教育和知识模块、文化和事件模块、品牌和对话模块、特别模块、旅馆和常住模块以及办事处和总部模块。

Zurich is the country's largest city and focal point of business, finance, higher education, and research. With 24,000 employees (2009), Zurich Airport serves 21.9 million passengers a year. Unique (Flughafen Zürich AG) has launched the development of a new major project at Zurich Airport. A high-quality building development for services with a utilizable area of around 200,000m2 is planned within walking distance of the terminal. It is called "The Circle at Zurich Airport". The project consists of seven modules closely linked: Health & Beauty, Education & Knowledge, Culture & Events, Brands & Dialogue, Special, Hotels & Long-stay and Offices & Headquarters.

入围 · finalist
UNStudio (建筑师事务所)
Ben van Berkel (建筑师)

团队 team: Astrid Piber · Nuno Almeida · René Wysk · Mirko Bergmann · Jörg Lonkwitz · Claudio Meletta · Sander Versluis · Tade Godbersen · Maud van Hees · Nanang Santoso · Kristina Madsen · Leon Bloemendaal · Patrick Noone

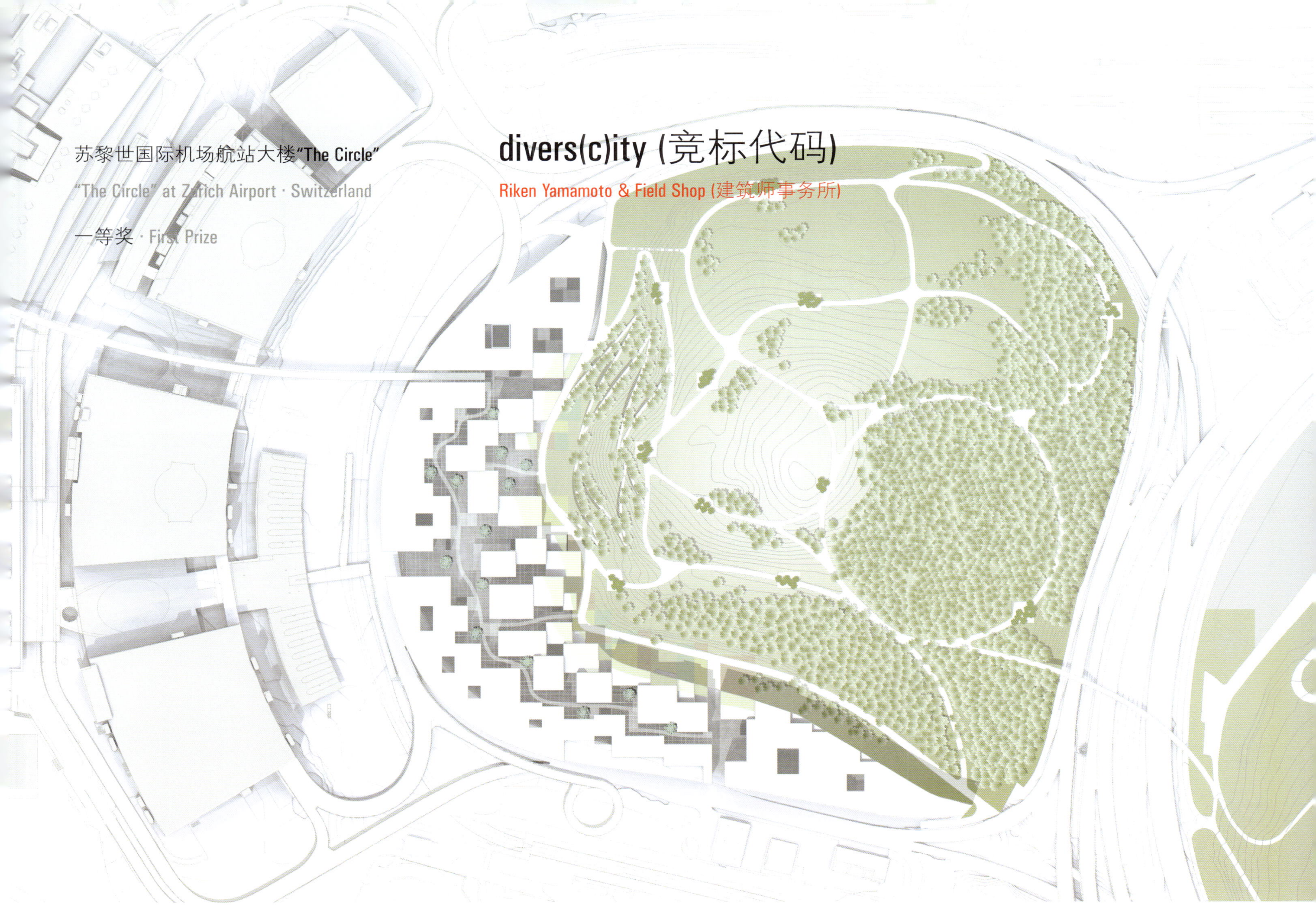

苏黎世国际机场航站大楼"The Circle"

"The Circle" at Zürich Airport · Switzerland

一等奖 · First Prize

divers(c)ity (竞标代码)

Riken Yamamoto & Field Shop (建筑师事务所)

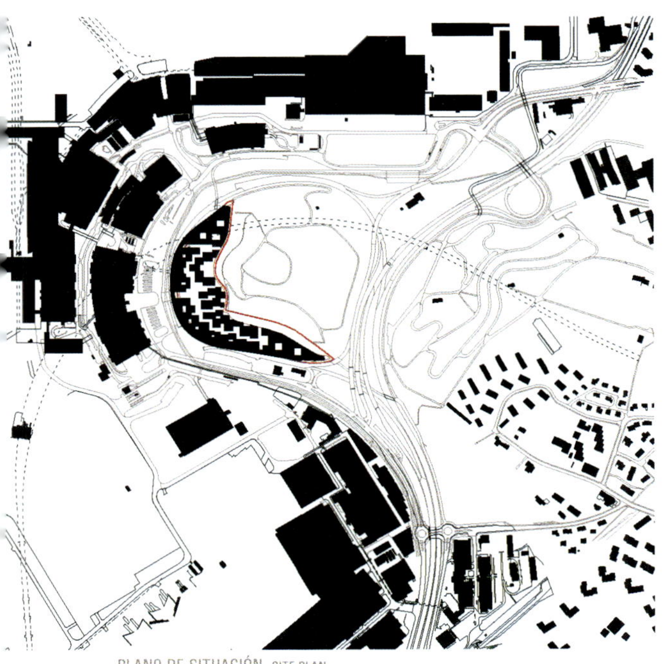

PLANO DE SITUACIÓN SITE PLAN

新型小镇

现代瑞士美在何处？我们来看看烙上瑞士印记的精密性。圆柱极其细长，使建筑看上去无比精致玲珑。建筑于中世纪时的瑞士城镇如今是人们日常生活的中心，而这些城镇因为完好保存了古老的建筑结构而被赋予了许多新的价值。它们的用途十分广泛且灵活多变。"The Circle"也是这样一个城镇。我们打算运用新科技、新建筑方法及设计手段来重建中世纪城镇的结构形态。从终端看，该工程是一处立面曲线幅度明显的巨型建筑。然而，从幽幽苍翠的山上俯瞰，它形如一座小城。

A NEW TOWN

What is the beauty of contemporary Switzerland? We want to show the Swiss precision. The columns, made extremely slender will produce architecture of unprecedented delicacy. The medieval cities of Switzerland that are the actual centers of everyday life for people today are extremely flexible because they retain their old structures. They can be used in an extremely flexible way. The Circle too is such a city. We would like to re-create the structure of a medieval city with new technology, construction methods and design. From the terminal, the project has the image of one enormous building with a dynamic curved facade. However, seen from the green hill, it seems like a small city.

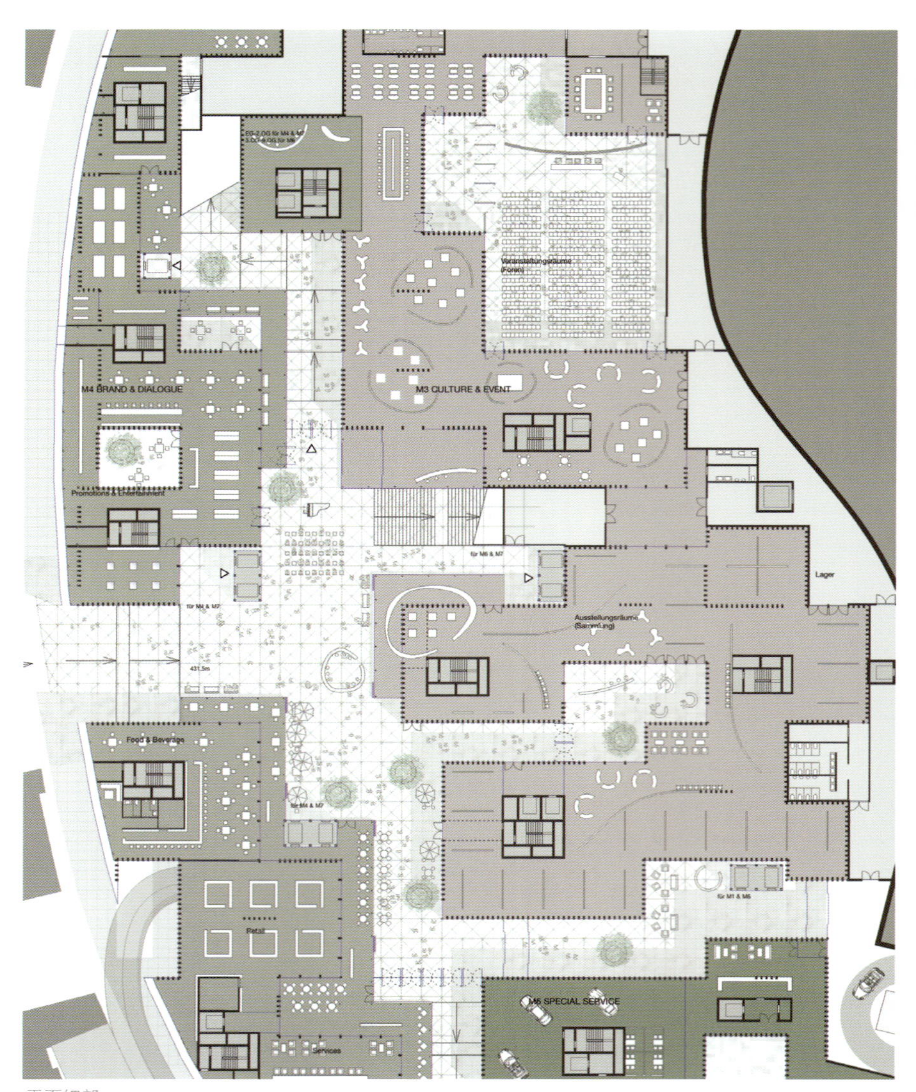

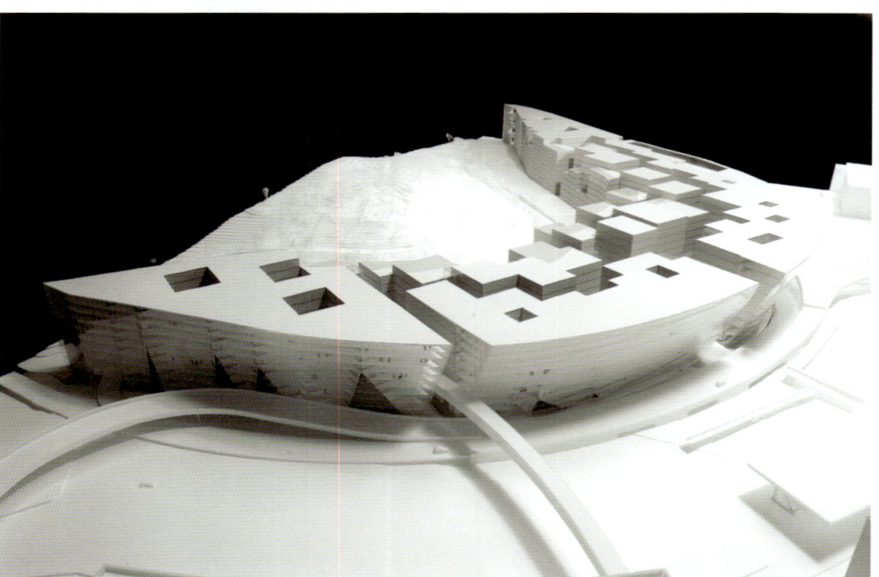

平面细部 DETAIL PLAN

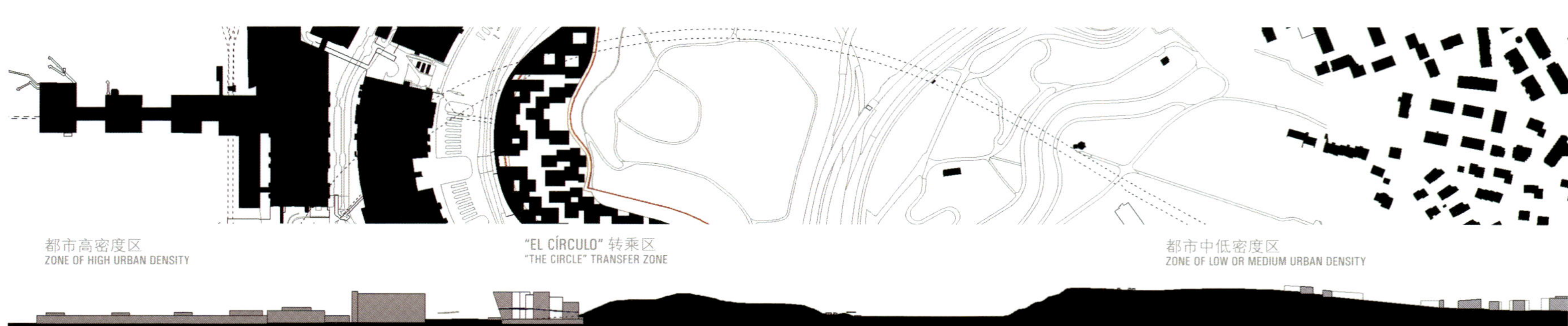

立面图 ELEVATION

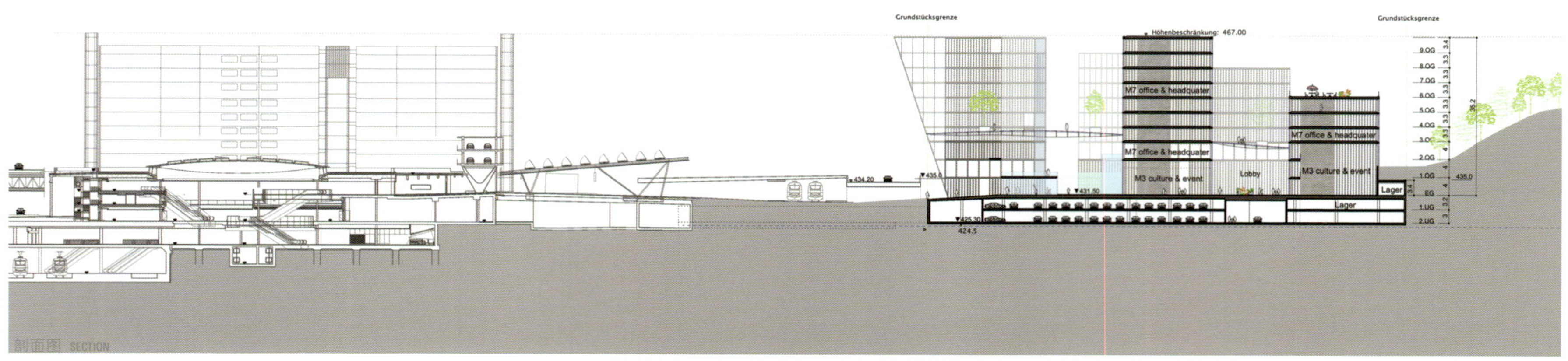

剖面图 SECTION

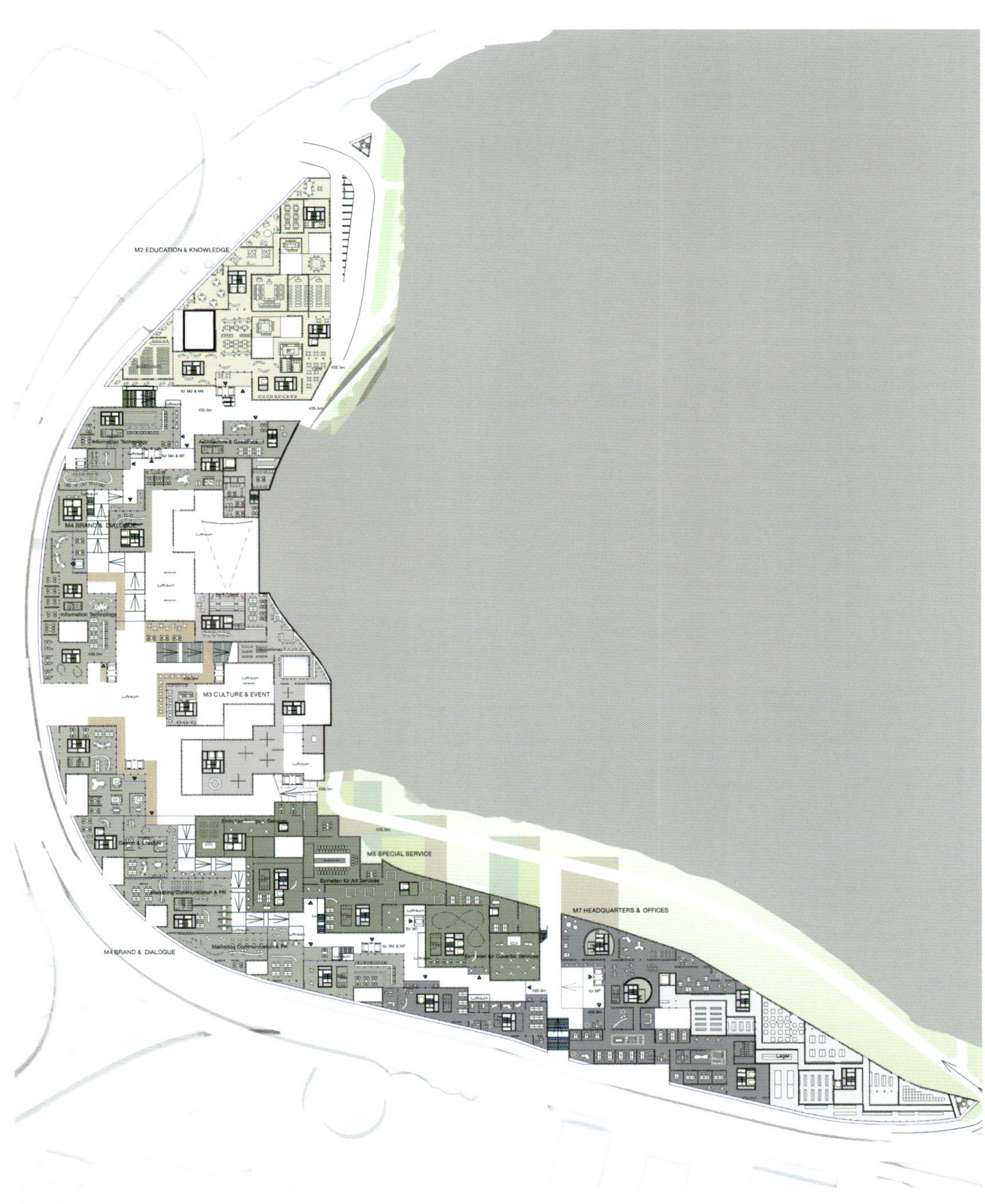

总平面图 OVERALL PLAN

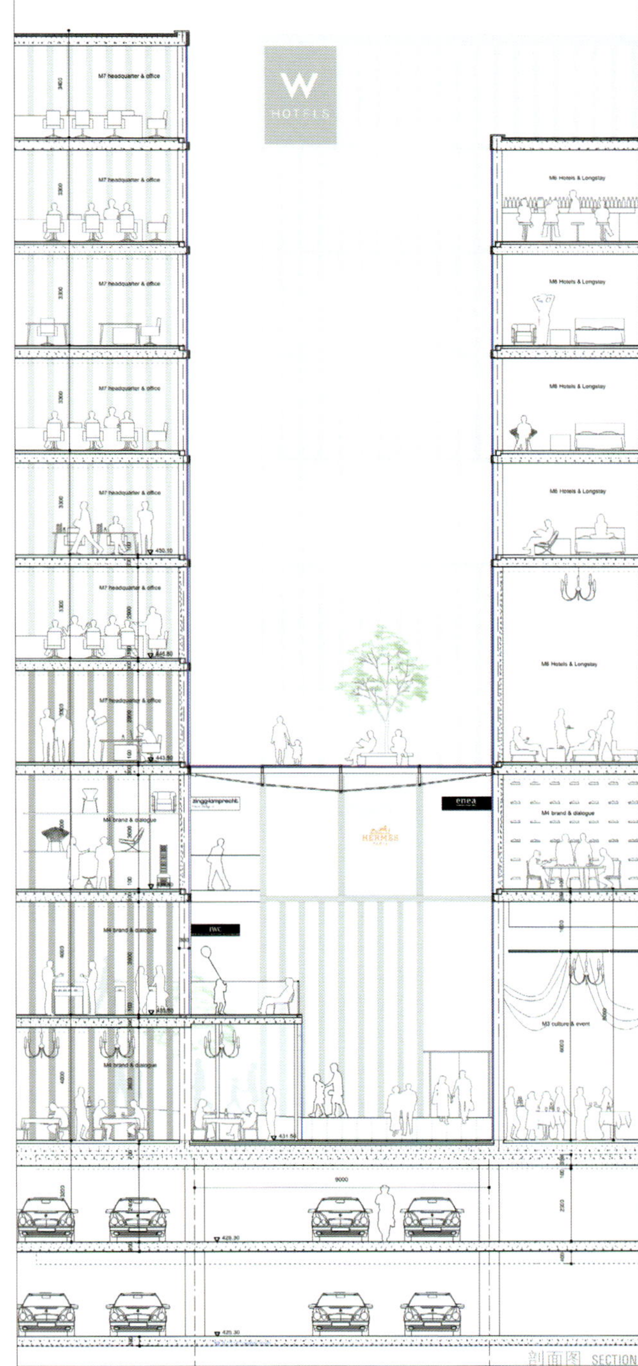

剖面图 SECTION

苏黎世国际机场航站大楼"The Circle"

"The Circle" at Zurich Airport · Switzerland

二等奖 · Second Prize

freitag (竞标代码)

Zaha Hadid Architects (建筑师事务所)

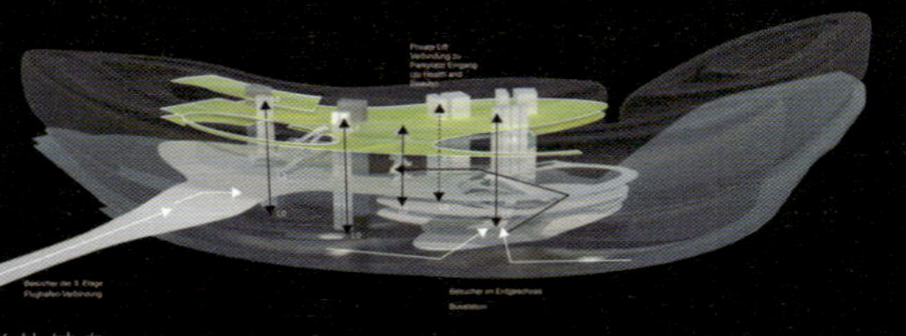

美体健身 HEALTH + BEAUTY

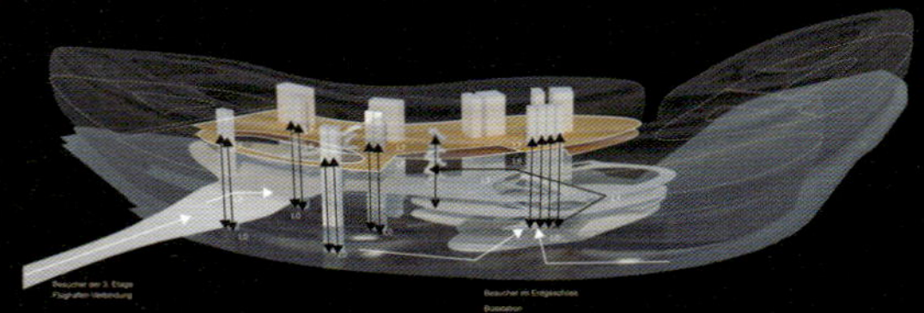

知识教育 EDUCATION + KNOWLEDGE

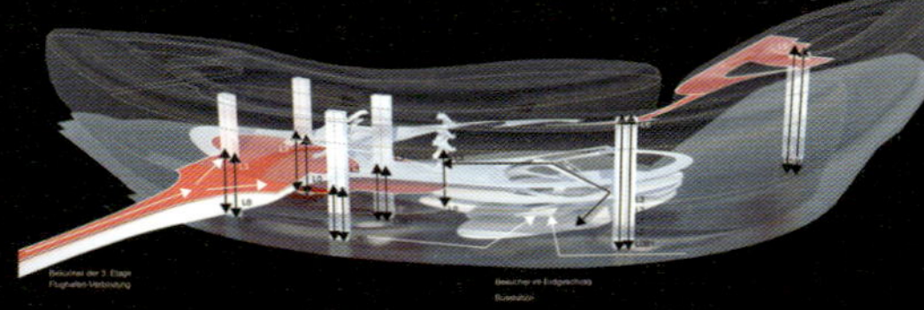

CULTURA + EVENTOS CULTURE + EVENTS

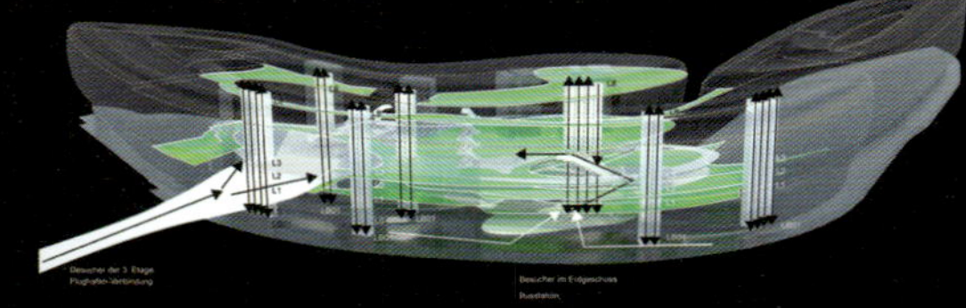

MARCAS + DIALOGO BRANDS + DIALOGUE

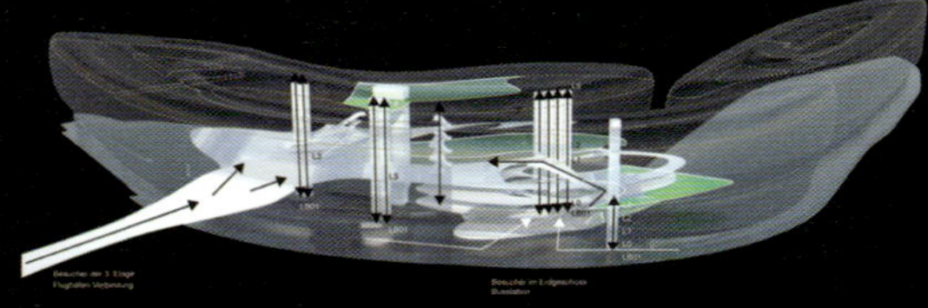

特殊服务 SPECIAL SERVICES

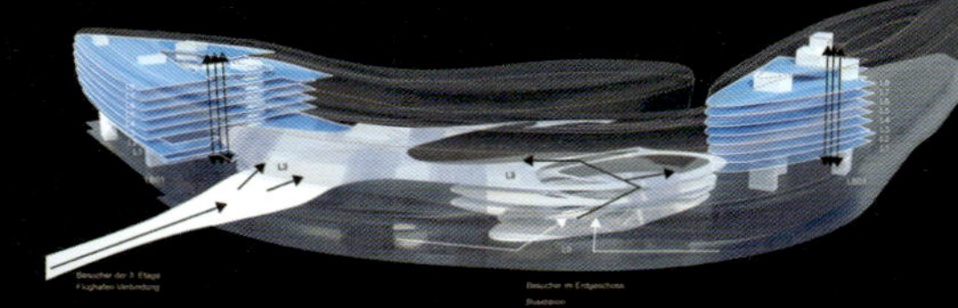

酒店长期停留 HOTEL + LONGSTAY

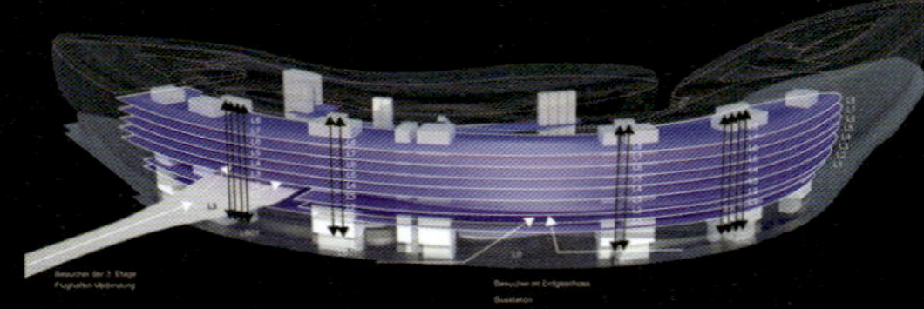

办公总部 1 HEADQUARTERS + OFFICES PHASE 1

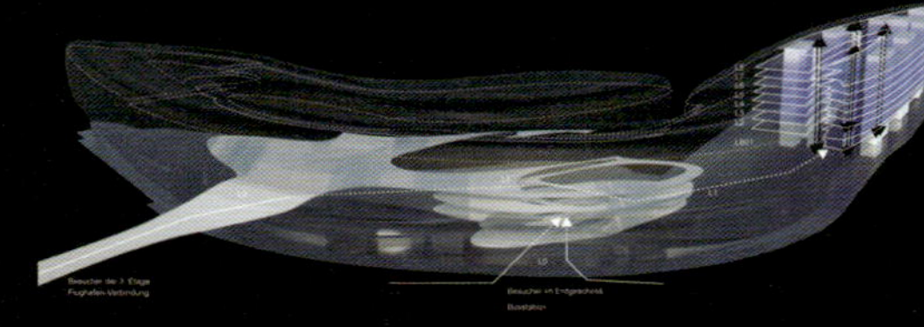

办公总部 2 HEADQUARTERS + OFFICES PHASE 2

清晰的通道

建筑的流体性需要具备两个条件：一面紧密不断地与机场相连，另一面凸起并经削减，这样就可以将周长最大化，足以环抱大自然并获得最佳视角。建筑的"内部都市特性"促进不同项目模块之间的互动，并将设备公开与公众共享，同时，可维持每个模块的实际垂直码放，从而使服务和循环变得更为合理。

CLARITY OF ACCESS

The fluid shapes of the building respond to two conditions. There is the compact and continuous adjacency to the airport on one side and a convex and eroded condition on the other side in order to maximise the perimeter, to embrace nature and to take maximum advantage of views. The building's "interior urbanism" encourages interaction amongst different programmatic modules and sharing of facilities open to the public, whilst maintaining a practical vertical stacking for each module, aimed at rationalizing services and circulation.

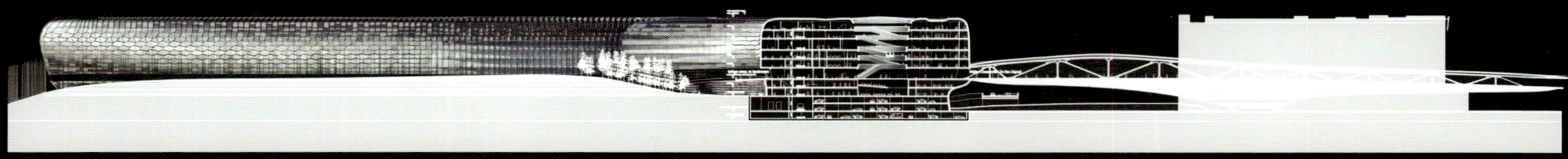

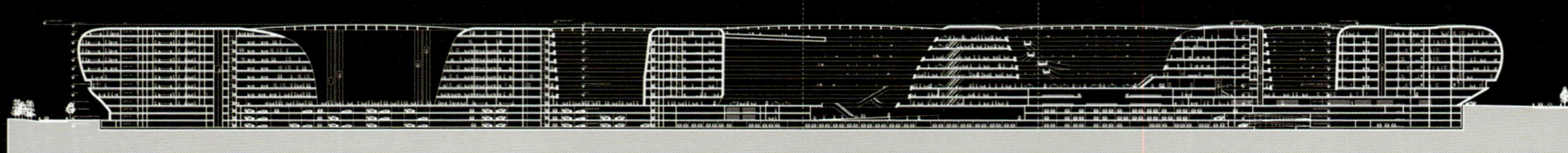

剖面图 SECTIONS

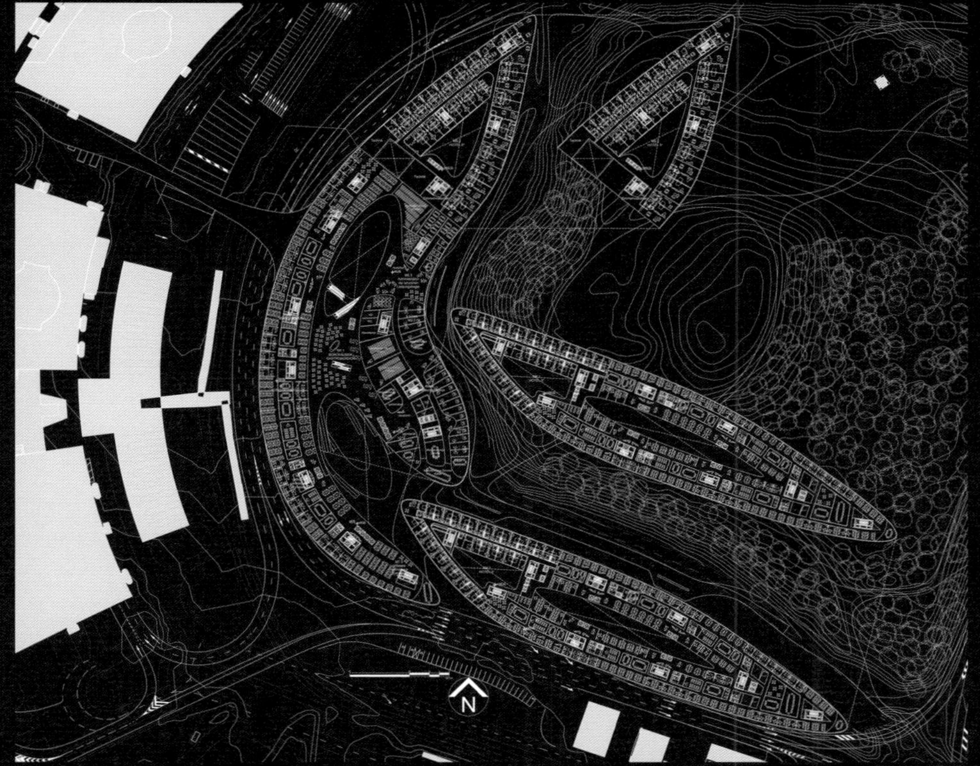

五层平面图 FLOOR PLAN 4

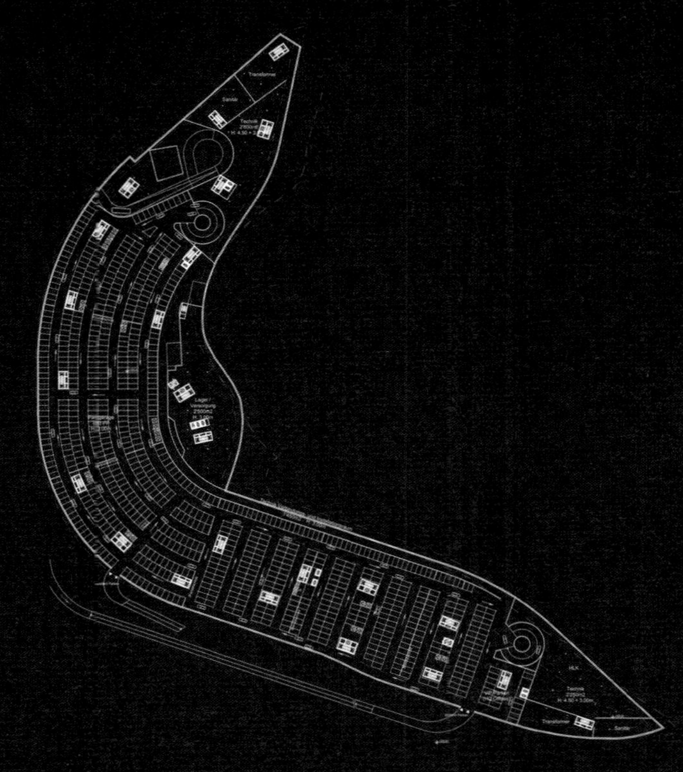

地下层平面图 BASEMENT FLOOR PLAN

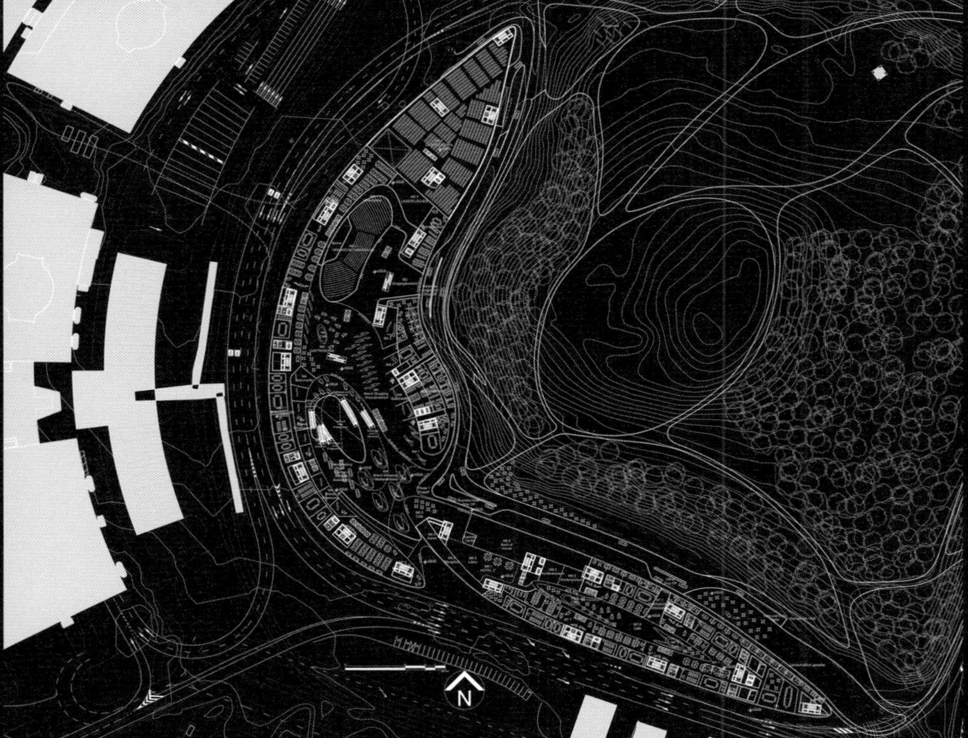

二层平面图 FLOOR PLAN 1

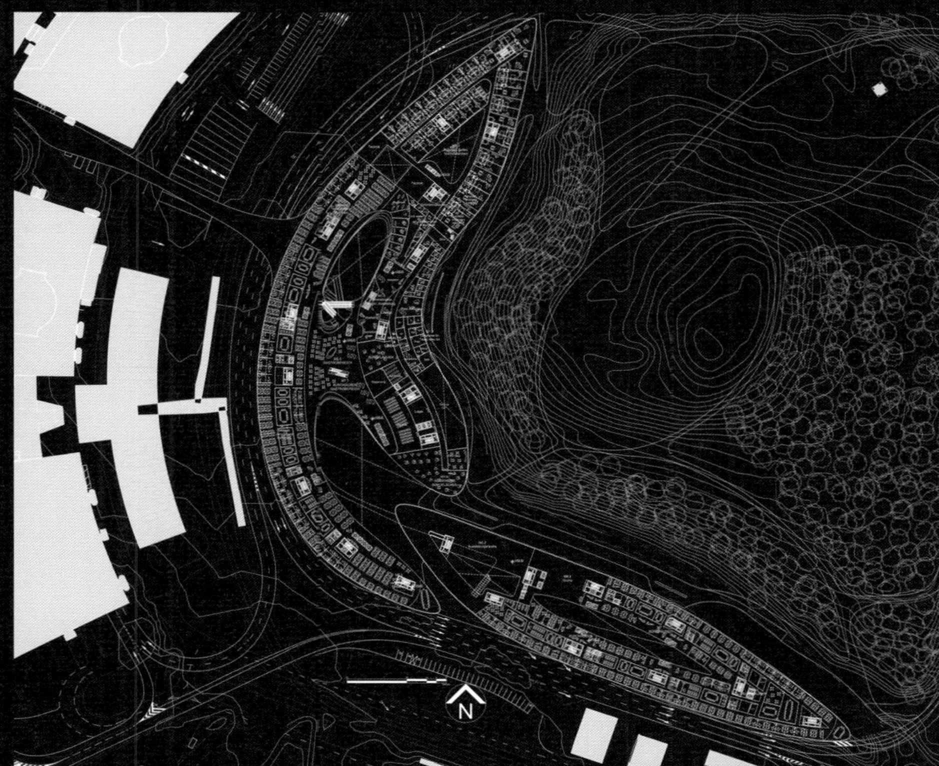

八层平面图 FLOOR PLAN 7

底层平面图 GROUND FLOOR PLAN

七层平面图 FLOOR PLAN 6

苏黎世国际机场航站大楼"The Circle"

"The Circle" at Zurich Airport · Switzerland

三等奖 · Third Prize

iskayska (竞标代码)

Xaveer De Geyter Architecten · Bureau Dan Budik (建筑师事务所)

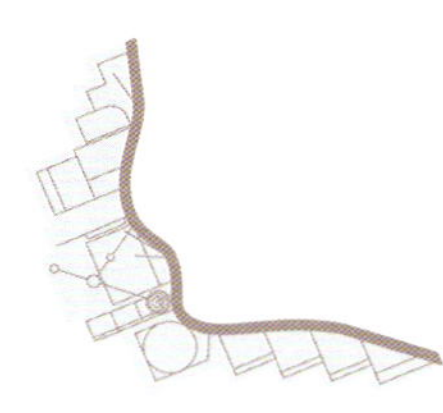

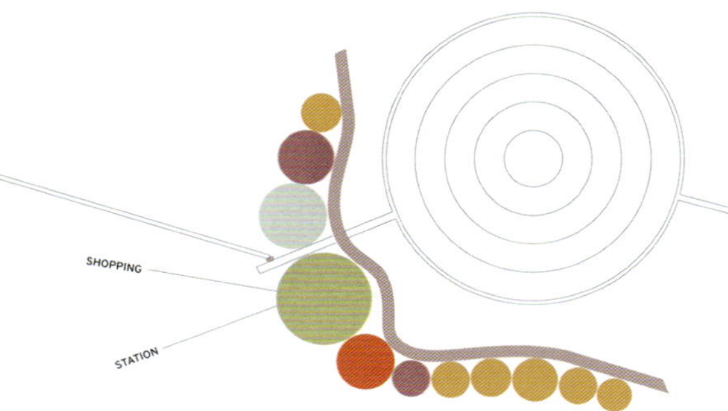

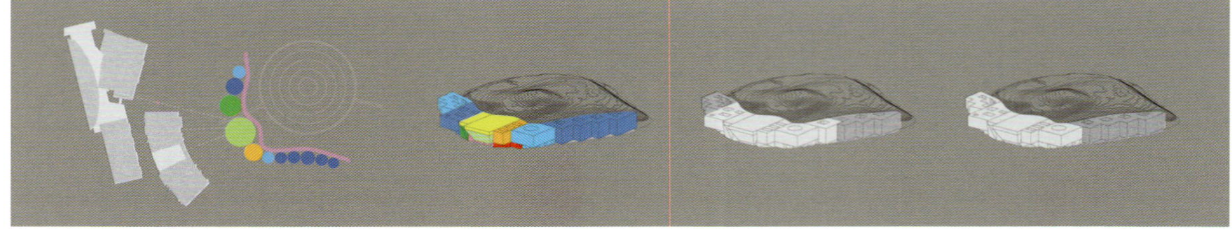

概念 CONCEPT

外立面图像 FAÇADE DIAGRAMS

不同的立面逻辑

我们首先决定将中央绿化带连为一体，并将其打造成这个新工程的象征性图标以及整个机场区的焦点。除了保留其自然元素之外，我们还加入了人工打造的新型公园元素。它们有序地排列成一连串同心圆，与当地瑞士山峦上具有天然层次的自然风貌相得益彰。这个新型的综合体有两个不同的立面，一个都市化的立面朝向机场，另一个更具田园风的立面朝向公园。在机场那一侧，很多建筑体交错矗立，表达了多样性和时代性的建筑诉求。这些建筑体均呈直角排列。而在公园那一侧，建筑表面则是一个巨大的平滑曲面，几乎完全延伸至机场边界。建筑立面由光亮材料打造而成，辉映着公园。

DIFFERENT FAÇADE LOGICS

Our first decision is to unify the central piece of greenery and to transform it into an icon for the new project and a central focus point for the whole airport area. While conserving its natural elements, new park ingredients of a more artificial nature are added. They are ordered through a series of concentric rings projected onto the existing topography, referring to the natural layering of nature on Swiss mountains. The new complex has two different faces, an urban one towards the airport, and a more Arcadian one towards the park. On the airport side a collection of volumes express diversity and simultaneity. They are all oriented orthogonally. On the park side, the appearance is defined by a gigantic smooth curve that almost entirely follows the site boundary. The shiny materials of the facade reflect the park.

面内环外立面 RING FAÇADE

面公园外立面 PARK FAÇADE

平面图 +431 FLOOR PLAN LEVEL +431

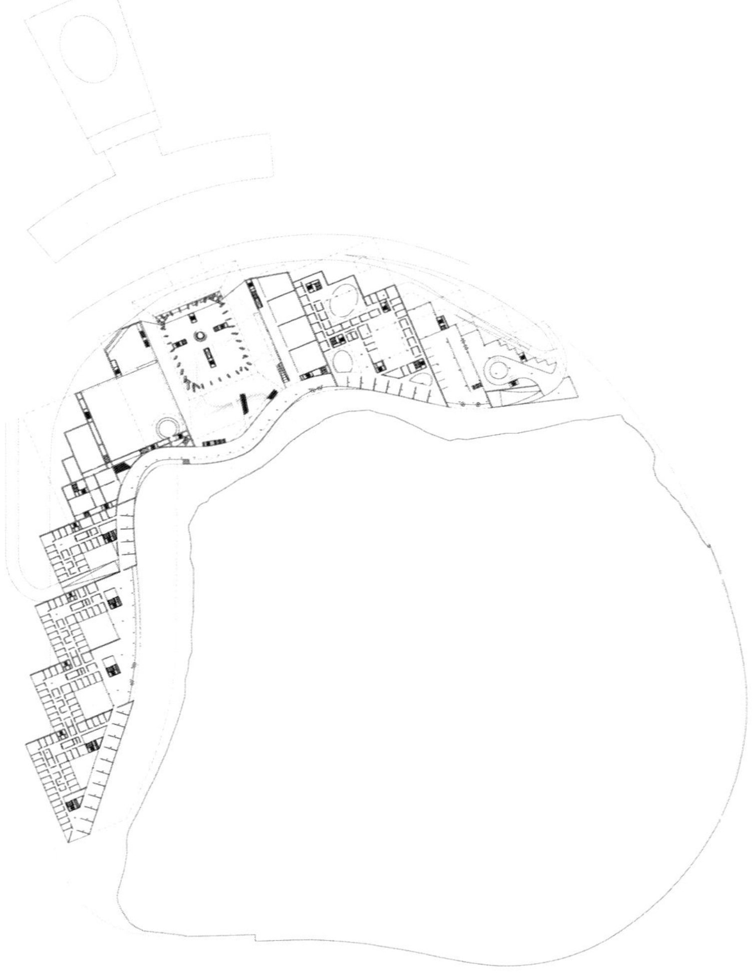

平面图 +433 FLOOR PLAN LEVEL +433

平面图 +442 FLOOR PLAN LEVEL +442

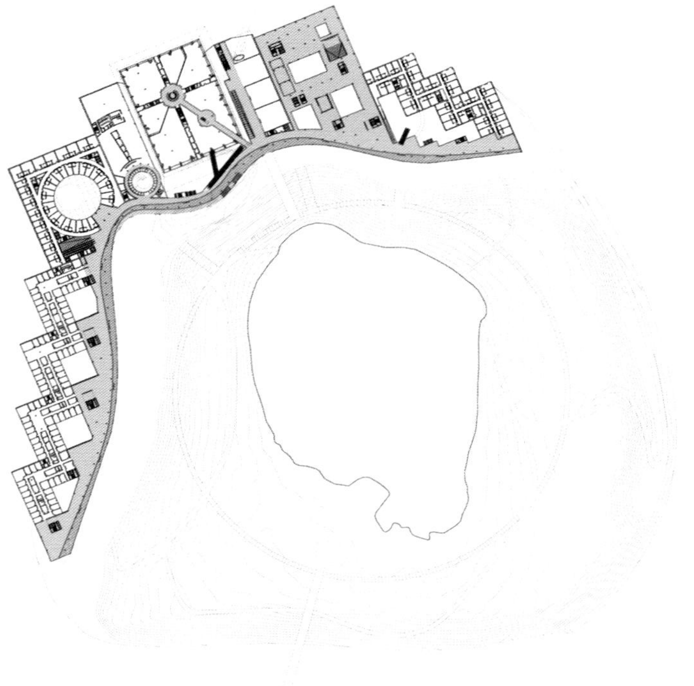

平面图 +447 FLOOR PLAN LEVEL +447

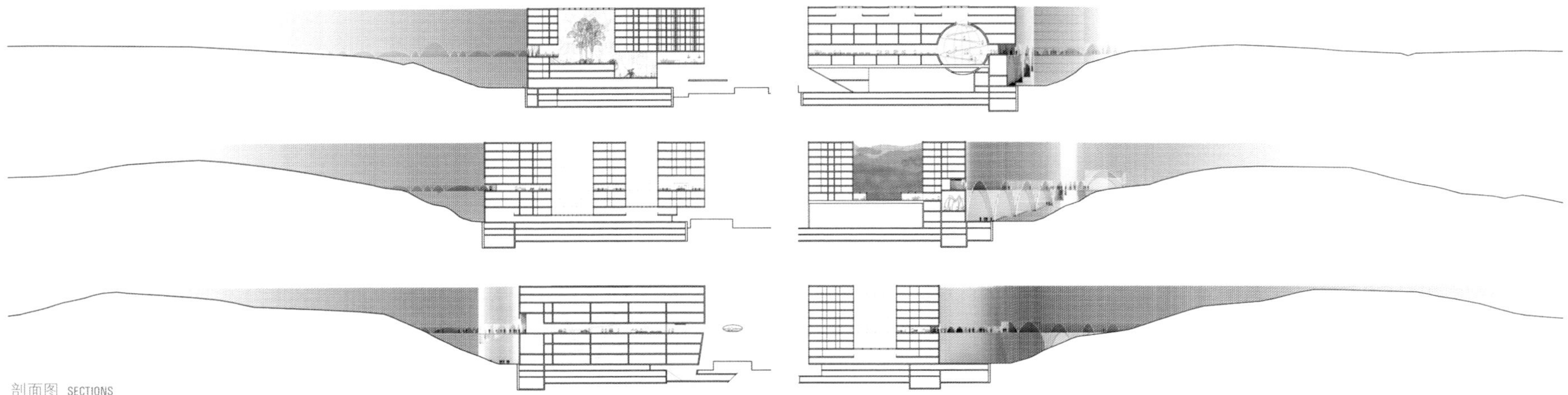

剖面图 SECTIONS

苏黎世国际机场航站大楼“The Circle”

“The Circle” at Zurich Airport · Switzerland

五等奖· Fifth Prize

halbmond (竞标代码)

Dürig AG (建筑师事务所)

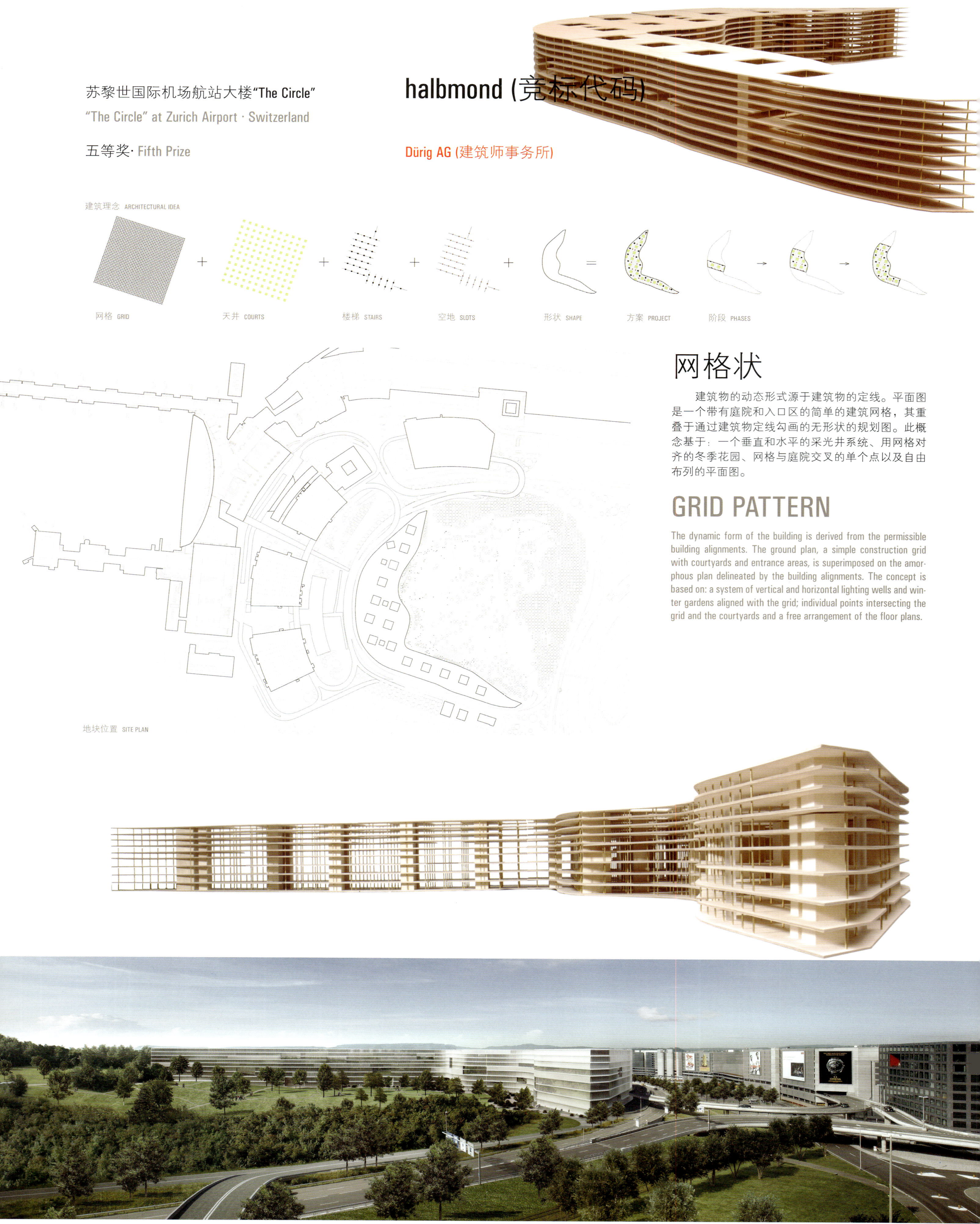

网格状

建筑物的动态形式源于建筑物的定线。平面图是一个带有庭院和入口区的简单的建筑网格，其重叠于通过建筑物定线勾画的无形状的规划图。此概念基于：一个垂直和水平的采光井系统、用网格对齐的冬季花园、网格与庭院交叉的单个点以及自由布列的平面图。

GRID PATTERN

The dynamic form of the building is derived from the permissible building alignments. The ground plan, a simple construction grid with courtyards and entrance areas, is superimposed on the amorphous plan delineated by the building alignments. The concept is based on: a system of vertical and horizontal lighting wells and winter gardens aligned with the grid; individual points intersecting the grid and the courtyards and a free arrangement of the floor plans.

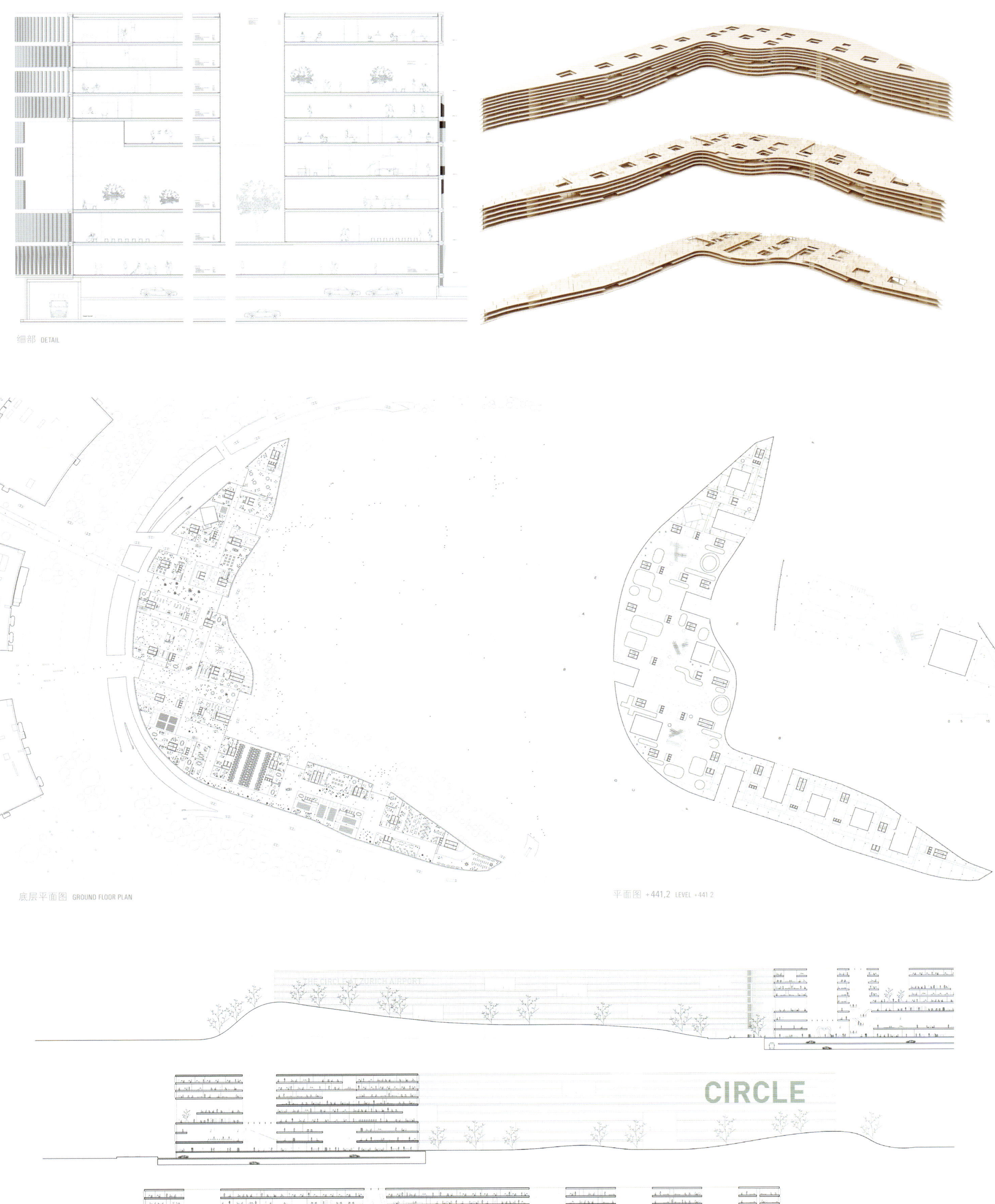

细部 DETAIL

底层平面图 GROUND FLOOR PLAN

平面图 +441,2 LEVEL +441 2

剖面图 SECTIONS

苏黎世国际机场航站大楼“The Circle”

“The Circle” at Zurich Airport · Switzerland

入围 · Finalist

allalin (竞标代码)

EM2N | MATHIAS MÜLLER | DANIEL NIGGLI | ARCHITEKTEN AG | ETH | SIA | BSA (建筑师事务所)

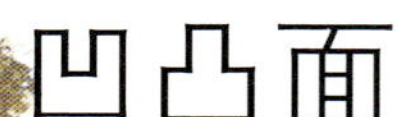

凹凸面

此建筑一面是面向城镇广场的大型凸形立面，另一面是位于Butzenbüel山脚下的凹形立面。此圆周就像一台过滤器位于内部和外部之间，并介于充满活力的城镇广场和Butzenbüel的绿洲之间。此建筑被分隔于多个层次之中，每层根据位置的不同而塑有不同的形状，同时其深度也会因此而不同。

CONCAVE AND CONVEX

The building develops a convex, large-scale face to the town square and a concave face, at the Foot of the Butzenbüel Hill. The Circle lies down like a filter between the inside and outside, between the lively town square and the green oasis of Butzenbüel. The building is organized in layers that are shaped differently depending on their location and create depth.

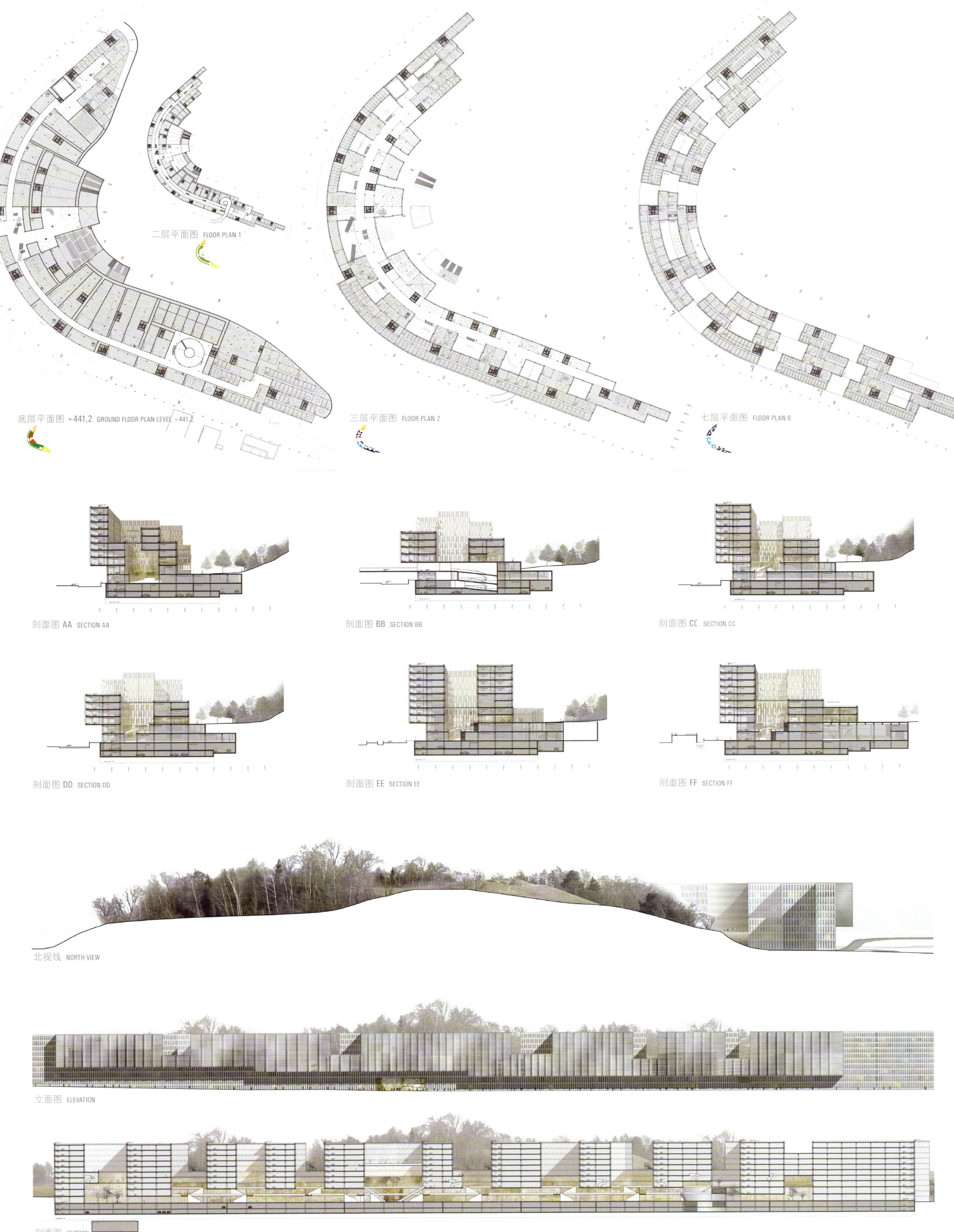
二层平面图 FLOOR PLAN 1
底层平面图 +441,2 GROUND FLOOR PLAN LEVEL +441,2
三层平面图 FLOOR PLAN 2
七层平面图 FLOOR PLAN 6
剖面图 AA SECTION AA
剖面图 BB SECTION BB
剖面图 CC SECTION CC
剖面图 DD SECTION DD
剖面图 EE SECTION EE
剖面图 FF SECTION FF
北视线 NORTH VIEW
立面图 ELEVATION
剖面图 SECTION

苏黎世国际机场航站大楼"The Circle"

"The Circle" at Zurich Airport · Switzerland

入围 · Finalist

zurich top room (竞标代码)

Cruz y Ortiz arquitectos (建筑师事务所)

Antonio Cruz Villalón · Antonio Ortiz García (建筑师)

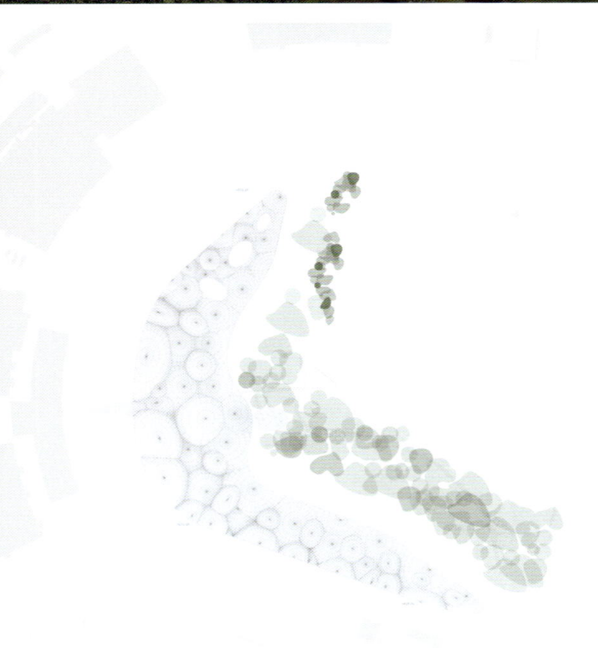

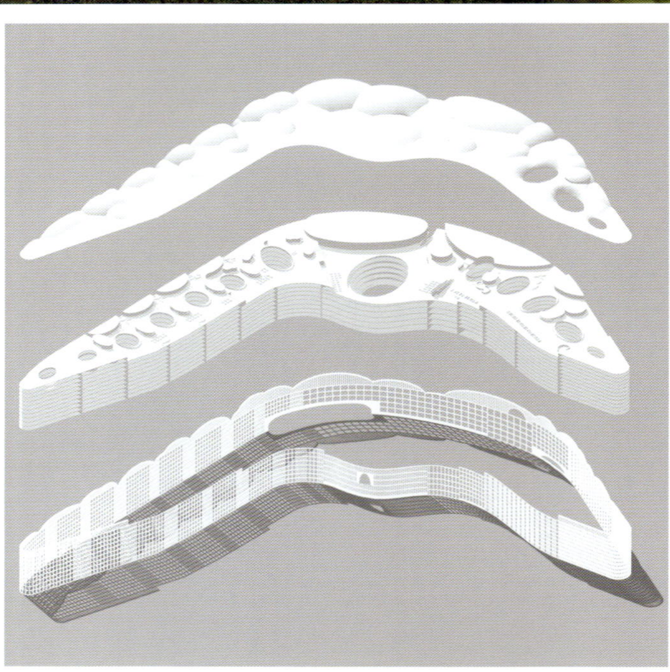

半透明与透明

我们提议建造一个带穹顶的大厅，不仅可以作为前台接待各方人士，而且还可以作为整个建筑的枢纽通往各个功能区。我们将这个空间称为"云端"。该建筑视野极佳，采光条件十分优越。

TRANSLUCID AND TRANSPARENT

We propose to create a great hall underneath the roof, a place to welcome, as a reception desk, with access to all modules. We call this space, "the clouds". It results in an excellent visibility and excellent light conditions.

地块位置 SITE PLAN

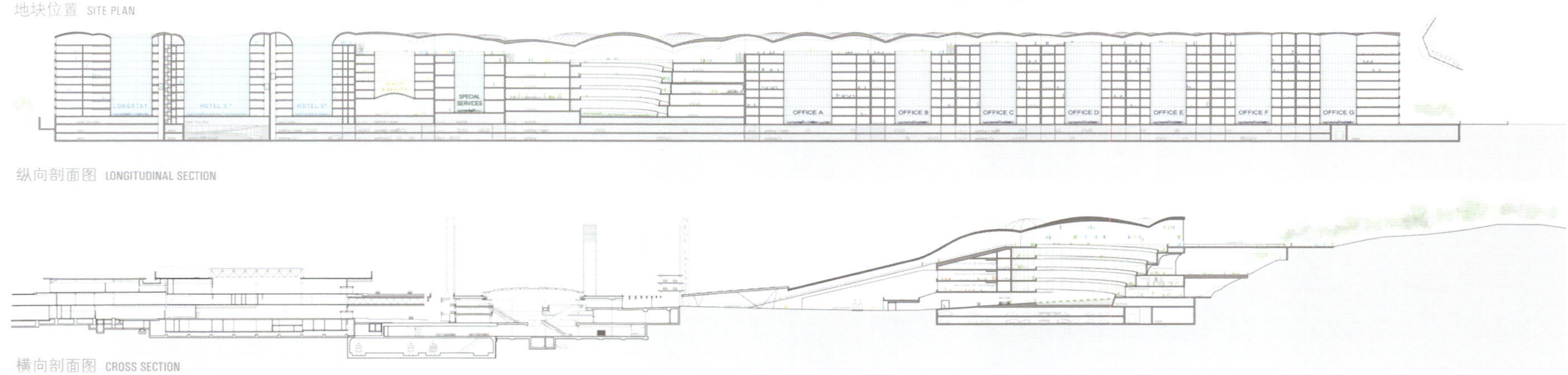

纵向剖面图 LONGITUDINAL SECTION

横向剖面图 CROSS SECTION

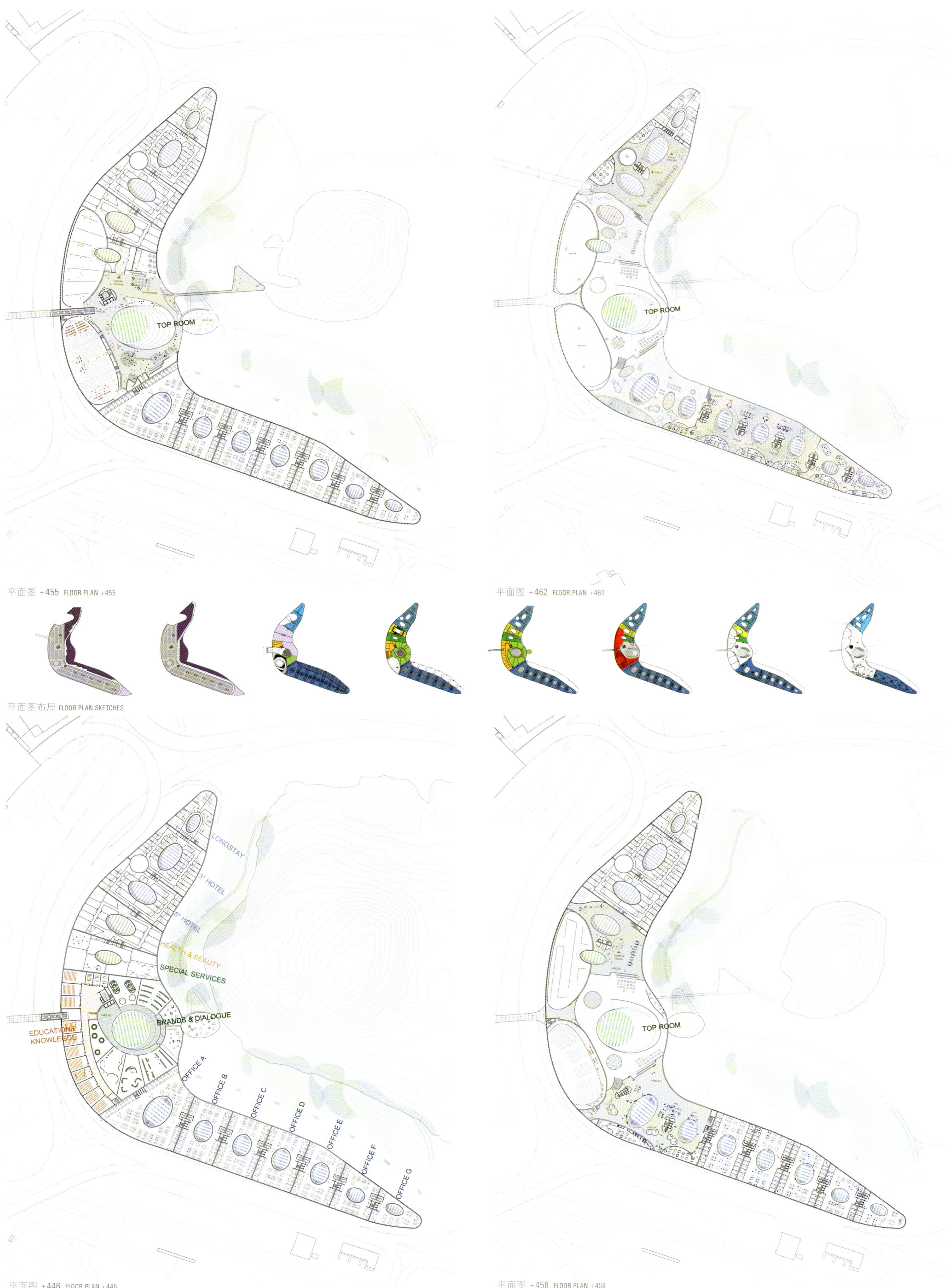

平面图 +455 FLOOR PLAN +455

平面图 +462 FLOOR PLAN +462

平面图布局 FLOOR PLAN SKETCHES

平面图 +446 FLOOR PLAN +446

平面图 +458 FLOOR PLAN +458

苏黎世国际机场航站大楼"The Circle"

"The Circle" at Zurich Airport · Switzerland

入围 · Finalist

z0252 (竞标代码)

UNStudio (建筑师事务所)

五个部分

此设计基于两个环境因素"机场"和"公园"，是类型学和几何学的综合体。此综合体构架由五个部分组成，与三角庭院相重叠，并向着公园的方向敞开，在两个都市区域之间创建了一个过渡区。重叠处倾斜且表面光滑，加上其中庭和正压送风系统，能够确保制冷、照明和通风所需的能量为最小。

FIVE PARTS

The design is based on a typological and geometrical amalgamation of two environmental factors 'airport' and 'park'. The complex is structured in five parts which together with overlapping triangular courtyards that open towards the park, creates a transition between the two urban worlds. The interaction of the inclined, and glazed façade, in addition to the atrium and positive pressure ventilation system, ensures that minimum energy is required for cooling, lighting and ventilation.

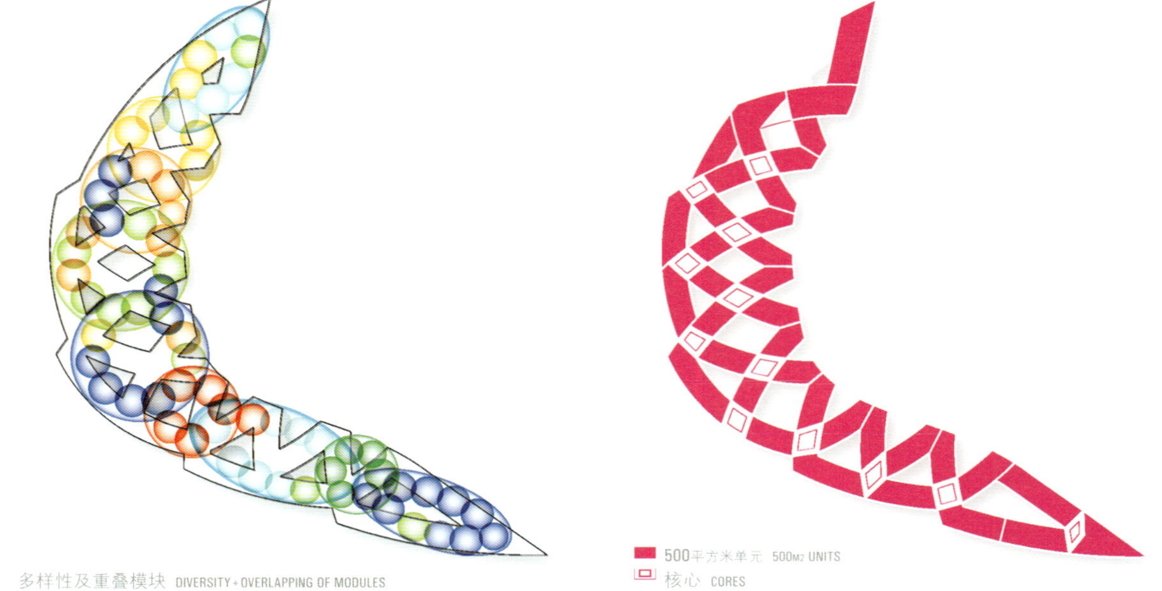

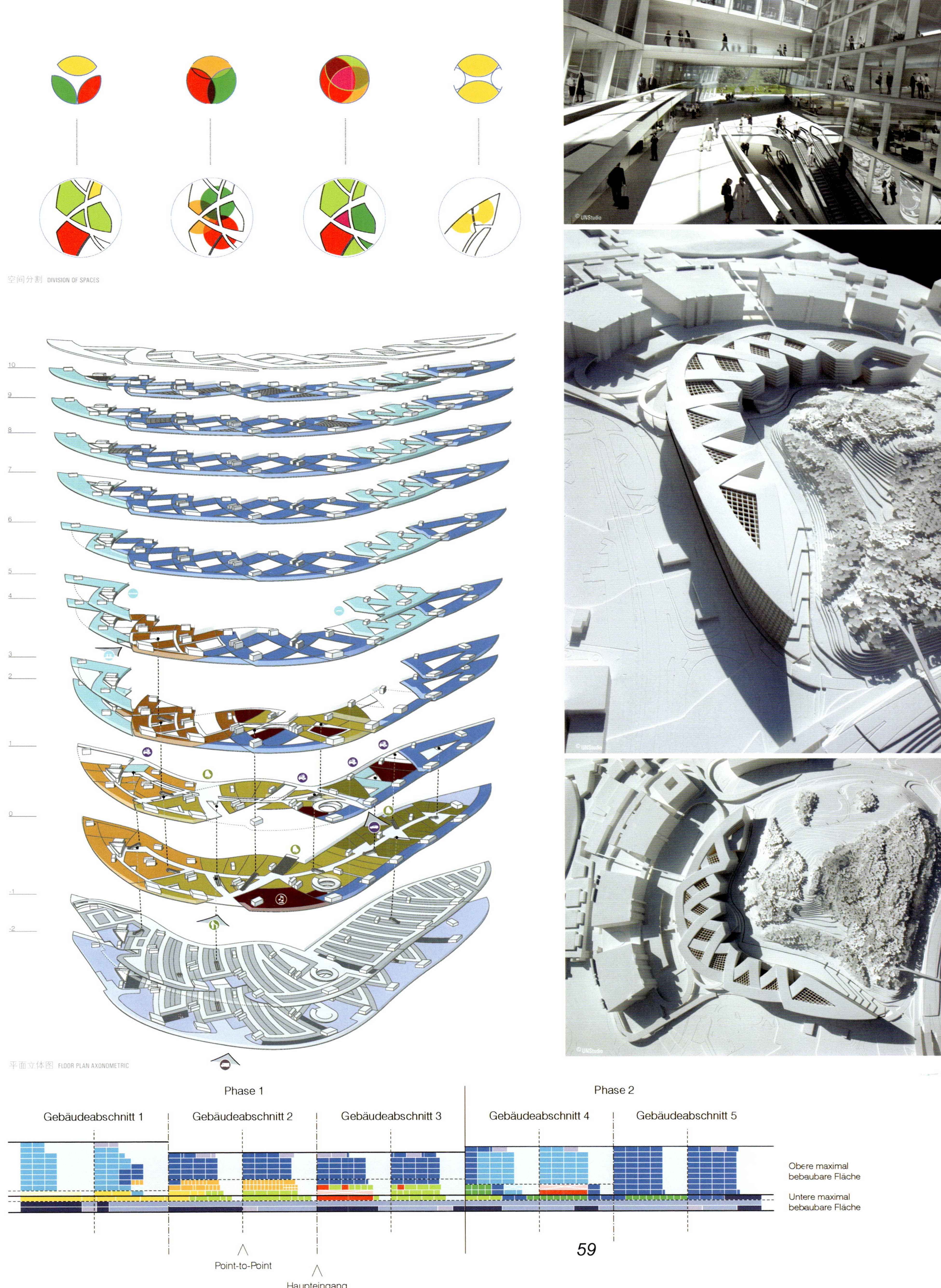
空间分割 DIVISION OF SPACES
10
9
8
7
6
5
4
3
2
1
0
-1
-2
平面立体图 FLOOR PLAN AXONOMETRIC
Phase 1
Phase 2
Gebäudeabschnitt 1
Gebäudeabschnitt 2
Gebäudeabschnitt 3
Gebäudeabschnitt 4
Gebäudeabschnitt 5
Obere maximal bebaubare Fläche
Untere maximal bebaubare Fläche
Point-to-Point
Haupteingang
© UNStudio

可持续都市交通中心 · 巴塞罗那

Sustainable Urban Transport Centre · Spain

竞标 · competition
可持续都市交通中心
Sustainable Urban Transport Centre

竞标类型 · competition type
concurso, abierto, internacional, en tres fases
three-stage open, international

项目地点 · site area
巴塞罗那 · 西班牙 Barcelona · Spain

主办方 · promoter
ArchMedium (organizes international architecture competitions for students)

日程安排 · schedule
招标 · Announcement 11.2009
评审结果 · Jury´s results 04.2010

评审团 · jury
Ignacio Paricio, Joyce de Botton
Zaida Muxí, Alfons Soldevila, Shaun Pilgrem

获奖者 · awards

一等奖 · first prize
BBEFK · 布宜诺斯艾利斯大学 Universidad de Buenos Aires
Bruno Bianchi · Damian Bojko · Nelson Etcheberry · Juan Fabbri · Guillermo Klentak
Christian Moroni

二等奖 · second prize
CIAN · FADU · 阿根廷圣达菲利托拉国立大学 Universidad Nacional del Litoral, Santa Fe, Argentina
Alfredo Gustavo Bertuzzi · Ileana Carolina Rossi Seluy · Nicolas Emanuel Sambrana

三等奖 · third prize
FF0o00
Sean Solowski · Peter Osborne · Farid Noufaily

提名奖 · honourable mention
Aleph
Beni Goltsman Barzellai · Guilherme de Barcellos Lozinsky · Mariana Magalhães Costa
Patricia Tinoco

提名奖 · honourable mention
EAD-Studio.net · 帕尔马大学 Università degli Studi di Parma
Daniel Valle Gómez · Flavio Panico

提名奖 · honourable mention
Estocolmo · 巴塞罗那高等建筑学院 Escola Tècnica Superior d'Arquitectura de Barcelona
Enric Navarro · Joan Ramon Pastor · David Onieva Alonso · Albert Sabás Serrallonga

提名奖 · honourable mention
GMT+2 · 阿尔卡拉高等建筑测量学院 Escuela Técnica Superior de Arquitectura y Geodesia de Alcalá de Henares
Sergio del Barco · Raúl Olivares · Pablo Magán · Guillermo Angulo · Marc Cabarga
María Lozano

提名奖 · honourable mention
Lam · 马德里高等建筑学院 Escuela Técnica Superior de Arquitectura de Madrid
Lu Lu · Antonio Berrio Gallego · Miguel Garcia Guerra

提名奖 · honourable mention
M+A · 克拉考工业大学 Cracow Univeristy of Technology
Marcin Giemza · Anna Skrzydlewska

提名奖 · honourable mention
Ortiz+Schvonewolff · 巴塞罗那高等建筑学院，代尔夫工业大学 ETSAB, TUDelf
Federico Ortiz · Ricardo Schvonewolff

提名奖 · honourable mention
Studio X · 肯塔基大学 University of Kentucky
Terry Allen Driggs, Jr.

提名奖 · honourable mention
Unoencinco · 布宜诺斯艾利斯大学 Universidad de Buenos Aires
Alejo Lopez · Noelia Maldonado · Laura Abate · Miguel Urruty · Maximo Triolo

入围 · finalist
Anxinal · 瓦耶斯高等建筑学院 Escola Tècnica Superior d'Arquitectura del Vallès
Iván Valero Fernández · Laura Domènech López

在Cerdà提出扩张巴塞罗那的宏伟计划150周年之际，该项目应运而生，建筑坐落在这个矩形网格中的一片区域，该区域是现代巴塞罗那都市化最早起步的地区之一。该地块现有两幢建筑物，均在当地家喻户晓。一个是当地的加油站，另一个是独家代理汽车品牌的汽车经销店。该项目建议建造一个**新型综合体，以此标志以汽油作为主要动力来源的汽车时代的结束。**旨在唤起民众的环保意识、教育民众关注环境的可持续性问题，在设计中将电动车引进城市，并采用技术手段帮助减少诸如光伏板、风力机等设备造成的污染。

Due to the 150th anniversary of Cerdà's master plan for the expansion of Barcelona, the project takes place on a site located in a consolidated sector of this rectangular grid that is considered one of the first examples of modern urbanism. On the plot there are currently 2 buildings, both very well known by the people of the area. One corresponds to a local gas station, while the other is a car dealership run by an exclusive car brand. The goal is to propose **a new complex that symbolizes the end of an era marked by gasoline-powered vehicles** and oil as the main power source. The project aims to aware and educated people on environmental sustainability issues, introducing in its design the imminent arrival of electric vehicles to the cities and technological elements that help to reduce pollution such as photovoltaic panels, wind mills, etc.

可持续都市交通中心·巴塞罗那

Sustainable Urban Transport Centre · Spain

一等奖 · First Prize

110061 (竞标代码)

BBEFK (团队)

Bruno Bianchi · Damian Bojko · Nelson Etcheberry · Juan Fabbri · Guillermo Klentak · Christian Moroni

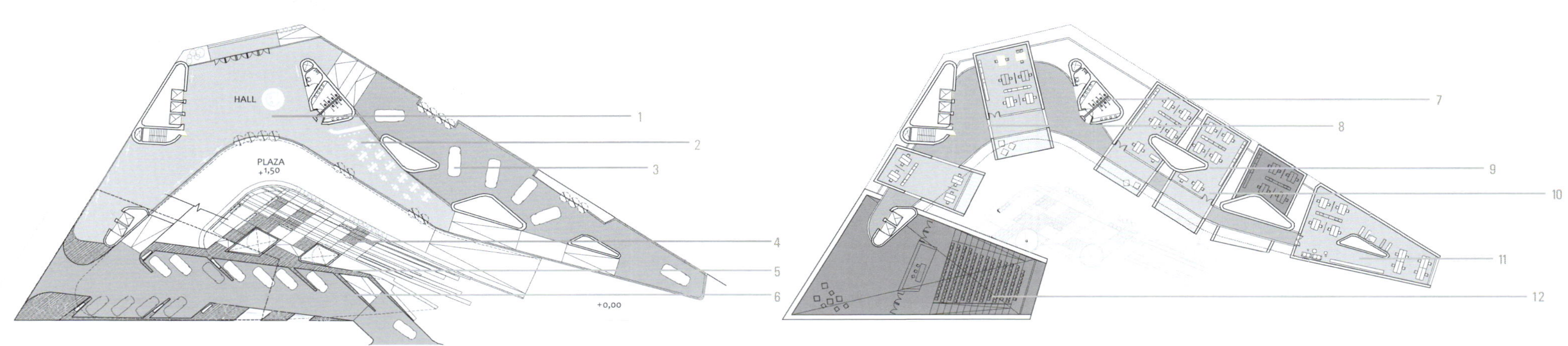

首层平面图 GROUND FLOOR PLAN

典型层平面图 TYPICAL FLOOR PLAN

1. 办公大厅 OFFICES' HALL
2. 咖啡 CAFETERIA
3. 经销商 CONCESSIONAIR
4. 花园广场 GARDEN SQUARE
5. 停车场入口 PARKING ENTRY
6. 停车场出口 PARKING EXIT
7. 西部站台-服务 WEST ROOT-SERVICES
8. 东部站台-垂直流通 EAST ROOT-VERTICAL CIRCULATION
9. 连接 INTERCONNECTOR
10. 办公室 OFFICES
11. 研究所 RESEARCH
12. 礼堂 AUDITORIUM

建筑广场

该项目旨在通过建造建筑广场来加强社交活动并与周边的绿化空间相映成趣。该建筑设计为城市的一个辅助系统，取意为一个由主建筑和不同背景构成的生命有机体，其中主建筑视野开阔，与周边环境共享盎然的绿意、充沛的雨水以及绝佳的空气流通。因此，该项目的内部结构是环环相扣，风格一致的。

BUILDING-PLAZA

The aim is to create a building-plaza, which promotes the social interaction and which locates in a system of green spaces. The building is conceived as a sub-system in the city, a living organism made up by a main structure open to the environment, where it obtains its energy, rainwater and ventilation, with a distribution to different settings. Thus the program is divided into inter-connected sectors within the system.

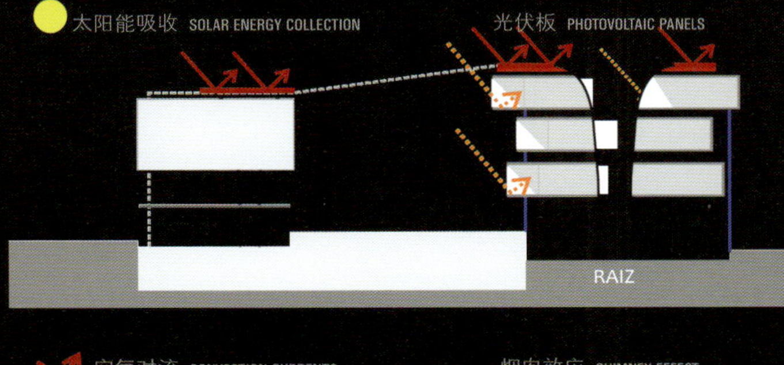

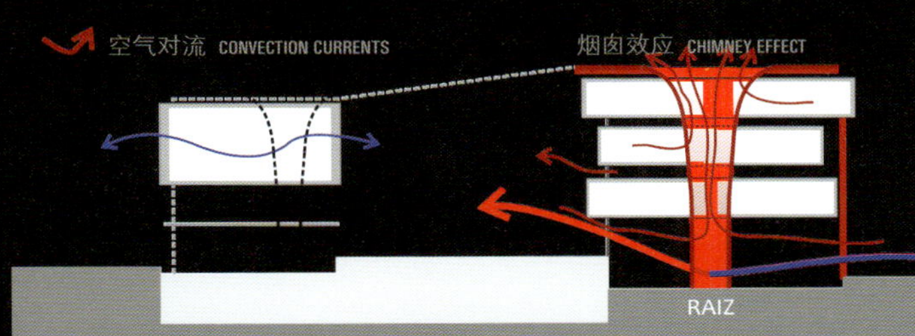

剖面图 SECTION

可持续都市交通中心 · 巴塞罗那

Sustainable Urban Transport Centre · Spain

二等奖 · Second Prize

110039 (竞标代码)

CIAN · FADU (团队)

Alfredo Gustavo Bertuzzi · Ileana Carolina Rossi Seluy · Nicolas Emanuel Sambrana

生态建筑

该建筑采用太阳电池板发电，可以满足建筑自身用电及充电中心的电力需求。其柱状结构体不仅起到支撑作用，而且将太阳能发电板转换的能量释放到底层，便于能量的可持续循环利用。外墙由钢和散热遮板建造而成，白天呈紫罗兰色，与太阳能电池板的颜色色调一致。太阳能电池板和散热遮板之间还安装了黄色、红色和蓝色的LED灯，使其在夜晚璀璨夺目。

AN ECOLOGICAL PLACE

The building uses solar panels to produce electricity to supply itself and supply the recharge centre. The columns that support it, not only have a structural function, they are also discharge points of the energy transformed by the solar panels which descend to the ground floor plan to the centre of sustainable transport. The volume shapes an outer skin of steel and louvers which maintain a violet colour during the day in reference to the solar panels. We pretend a delicate lighting of this big piece during the night. Yellow, red and blue LED lights are located between the solar panels and the louvers.

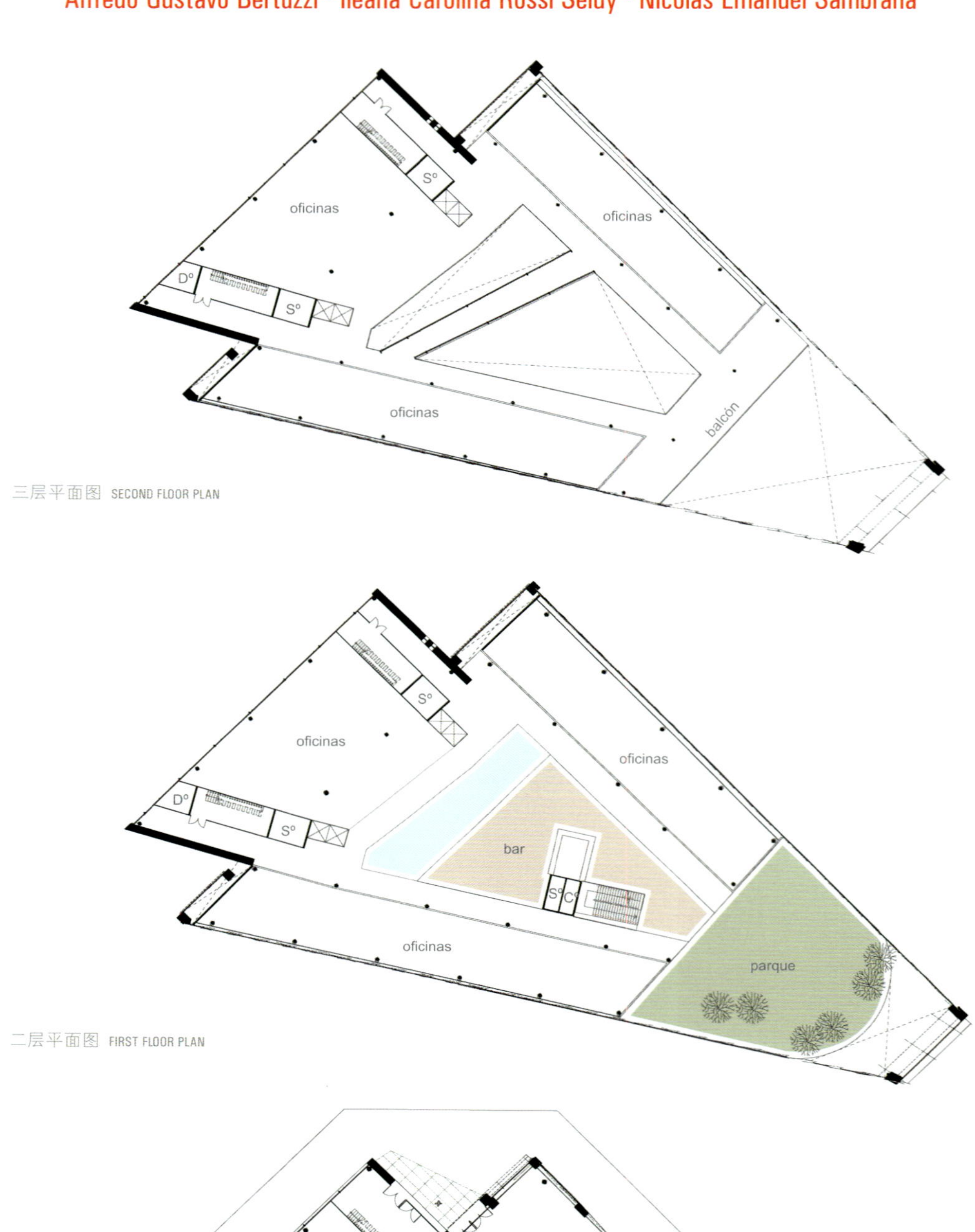

三层平面图 SECOND FLOOR PLAN

二层平面图 FIRST FLOOR PLAN

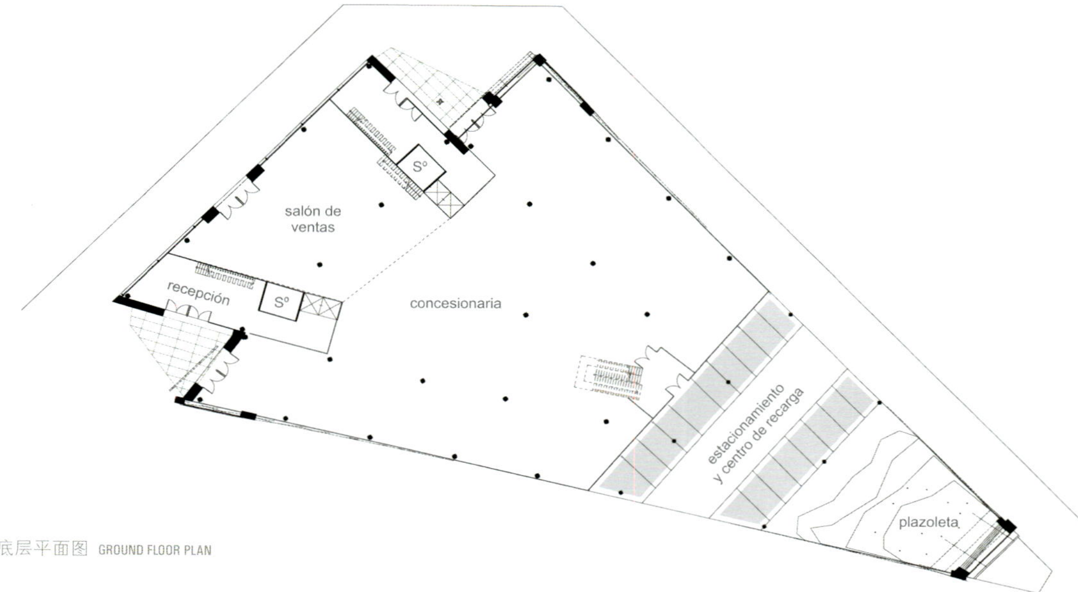

底层平面图 GROUND FLOOR PLAN

可持续都市交通中心 · 巴塞罗那

Sustainable Urban Transport Centre · Spain

三等奖 · Third Prize

130326 (竞标代码)

ff0d00 (团队)

Sean Solowski · Peter Osborne · Farid Noufaily

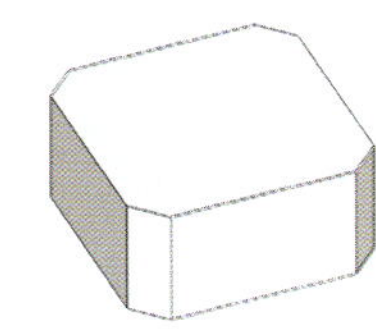

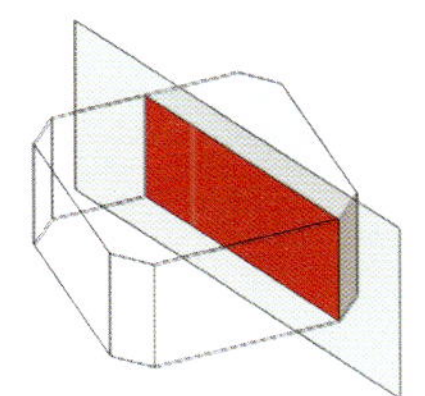

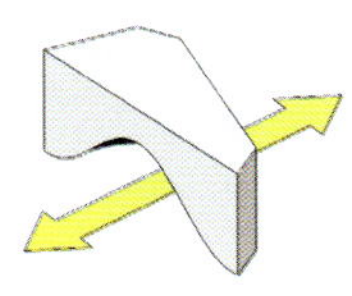

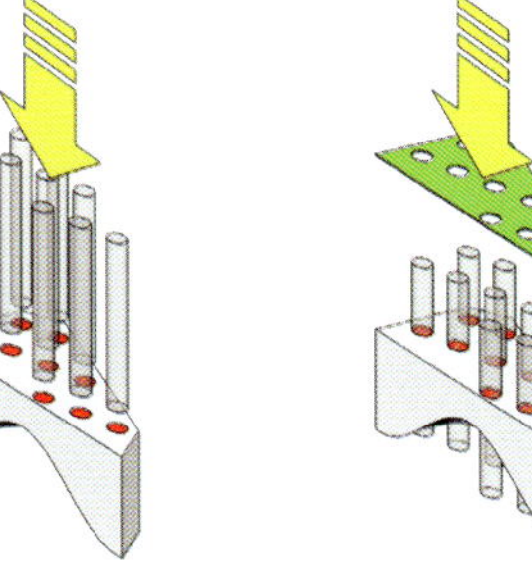

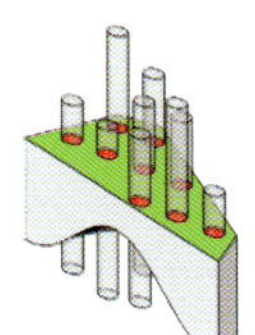

椭圆体

Castell de Cotxes是一个包含了运输体系改良构架的新型结构体，目的在于减轻我们对环境施加的影响。这是一个集绿化空间、电动车充电站、公共雕塑、零售便利设施以及研究型实验室为一体的多层次结构体，实现了土地的最大化利用，打造了一个社区服务中心，并且为环境研究和绿化延伸提供了便利条件。

为了教育大众和鼓励可持续技术的发展，该项目设计了16个圆柱体给料塔，每小时可为200多辆车辆服务，容量超过一个普通加油站的容量。该项目可设计在城市的任何闲置场所，也可设计在加油站所处的地段。凸起的台基和周边的各种公共服务设施相融为一体。

ELLIPTICAL CYLINDERS

Castell de Cotxes is a new facility that embraces an evolving framework of transportation systems to soften our impact on the environment. Combining green space, an electric car charging station, public sculpture, retail amenities and a research laboratory, the multi-level facility maximizes land-use, creating a community and service hub, while supporting environmental research and green transit.

To educate the public and celebrate sustainable technologies, a series of 16 cylindrical charging towers are designed to serve over 200 vehicles per hour, exceeding the capacity of a conventional gas fueling station. Our design can be implemented throughout the city on residual lots or where current gas stations exist. The raised podium integrates a myriad of public services.

LED Advertising Billboard
Solar Voltaic Panels
Sales Towers
52 cars
Rental Towers
130 cars
Charging Towers
High Effecient Light Wells / Skylights (Solatube)
Storm Water Retention
Hard Surface Gather and Recreation Area
Skylights with Integrated Passive Cooling Ventilation
Adaptive Louver Solar Shading + Privacy System
Floor Plan Program Strategy
Cafeteria Upper Deck
Electric Car Dealership
Cafeteria / Bar
Internal Circulation Ramps
LED Lighting Integrated into Reflective Ceiling Lighting System
Secondary Circulation Ramp
(Each servicing Independent Program) 3x Elevator Cores
(Each servicing Independent Program) 3x Stair Cores
Geothermal Wells Integrated intoVehicle Elevator Shafts

CHARGING TOWERS TORRE ELECTRICA
PUBLIC PARK PARQUE PUBLICO
BUILDING ENVELOPE ENVOLVENTE DE LAS EDIFICACIONES
BUILDING FLOOR PLATES NIVELES DE CONSTRUCCIÓN
UNDER CARRIAGE POR DEBAJO DE
GRADE TIERRA

可持续都市交通中心 · 巴塞罗那

Sustainable Urban Transport Centre · Spain

提名奖 · Honourable Mention

110097 (竞标代码)

Aleph (团队)

Beni Goltsman Barzellai

Guilherme de Barcellos Lozinsky

Mariana Magalhães · Costa Patricia Tinoco

建筑体的重新阐释

起源于150年前的Cerda计划旨在重新阐释建筑体的类型。该项目强调对地块的充分利用，使生活和交通空间沿着宽敞的中央空间周边紧密排列。研究中心成为了城市的延续空间。公共空间渗透到了建筑体内部。内部空间和外部空间之间的连续体是象征城市动态发展的综合体。坡道设计将城市的动态发展具象化，而延伸的流畅空间则将中心里不同的活动连为一体。

位置 LOCATION

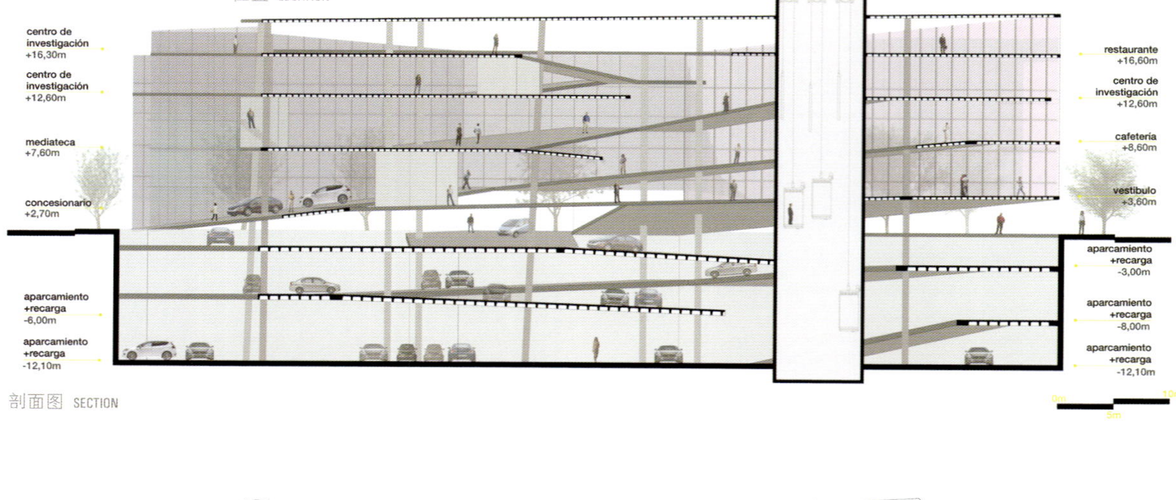

剖面图 SECTION

A REINTERPRETATION OF THE BLOCK

In virtue of the plan of Cerda created 150 years ago, the objective is to promote a reinterpretation of the typology of the block. The project promotes the full occupancy of the plot, concentrating the living and circulation spaces in the perimeter, which surround a big central space. The research centre works as an extension of the city. The public space penetrates inside the building. The continuity between the inside and the outside is a synthesis of the movement of the city. The ramps are the formalization of this movement; a fluid space which extends and allows the connection between the different activities of the centre.

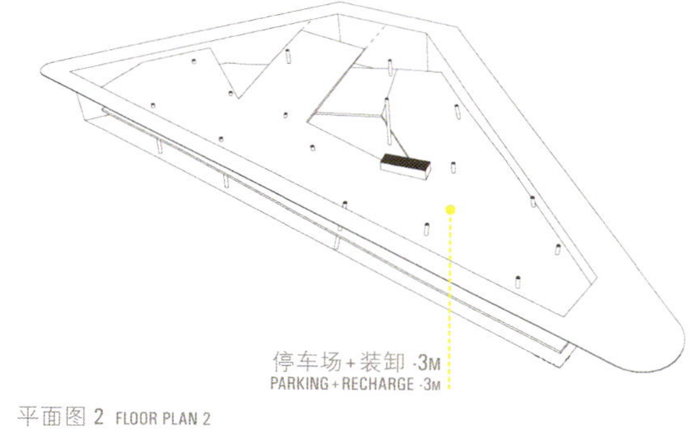

平面图 2 FLOOR PLAN 2

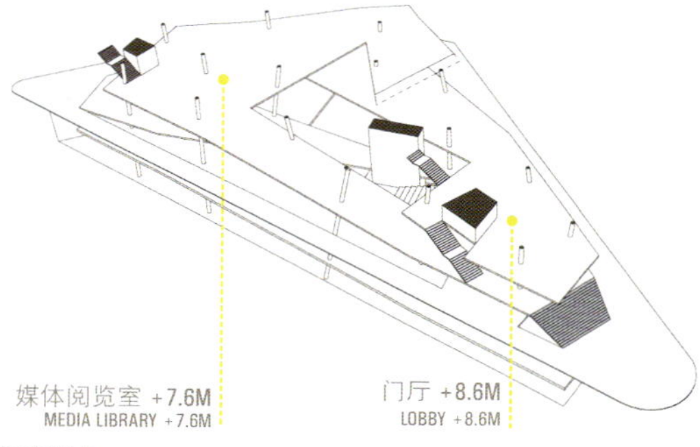

平面图 4 FLOOR PLAN 4

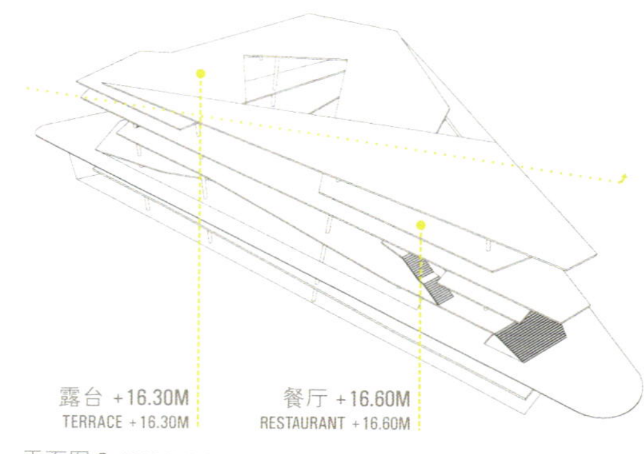

平面图 6 FLOOR PLAN 6

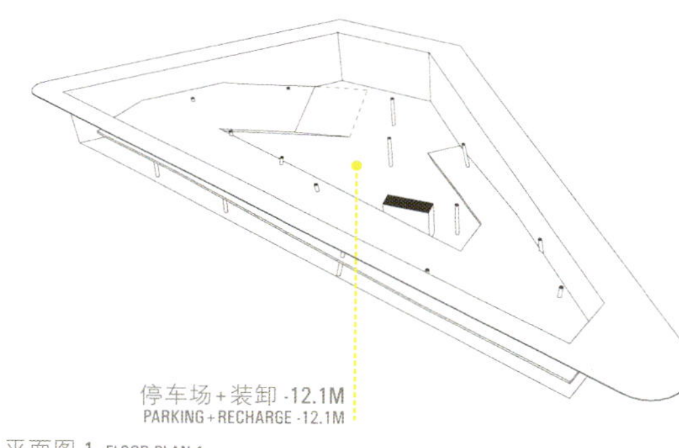

平面图 1 FLOOR PLAN 1

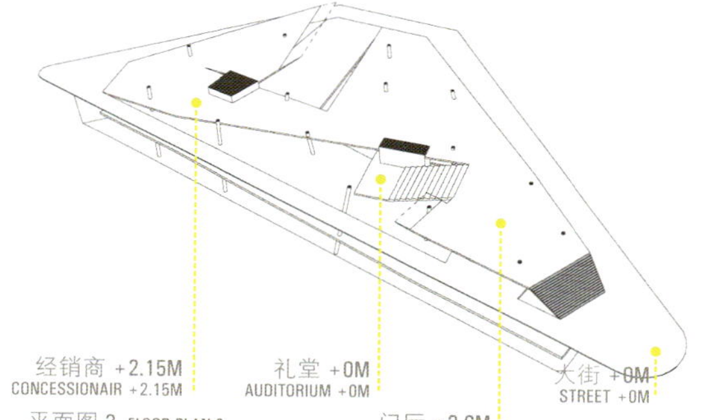

平面图 3 FLOOR PLAN 3

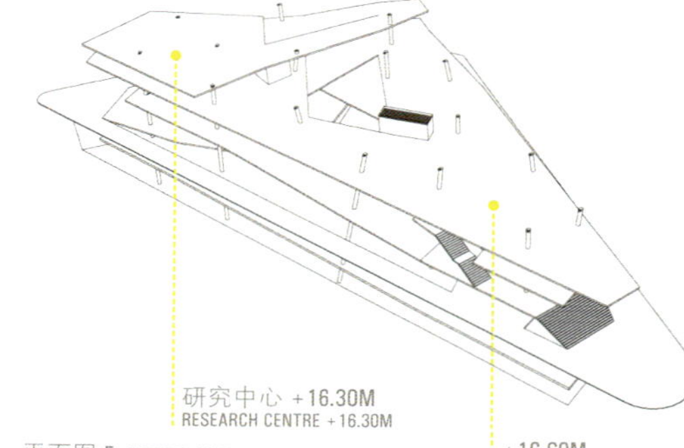

平面图 5 FLOOR PLAN 5

可持续都市交通中心 · 巴塞罗那

Sustainable Urban Transport Centre · Spain

提名奖 · Honourable Mention

110023 (竞标代码)

EAD-studio.net (团队)

Daniel Valle Gómez · Flavio Panico

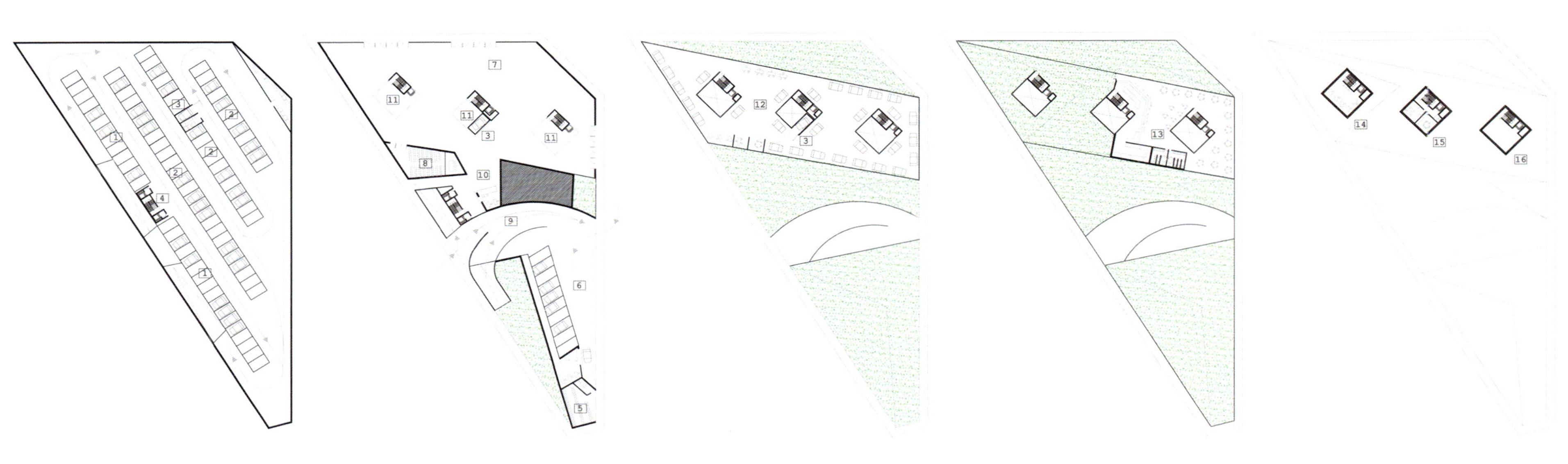

平面图 -1 FLOOR PLAN -1　平面图 0 FLOOR PLAN 0　平面图 1 FLOOR PLAN 1　平面图 2 FLOOR PLAN 2　典型层平面图 TYPICAL FLOOR PLAN

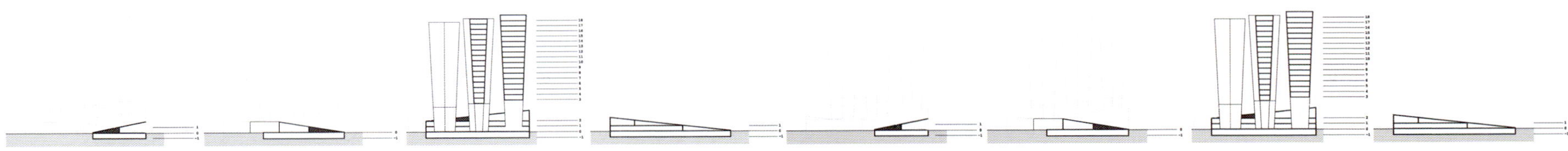

剖面图 SECTIONS

三座大厦

该项目分别坐落于两个不同的区域，其中一侧是低层建筑及地下室，另一侧则耸立着三座又细又高形似烟囱的大厦，采用的是嵌花式的设计手法。项目大部分的建筑正坐落于此。

低层建筑区域是建造在三个斜面上的一个巨大平面，打造了一个开阔宽敞的公共空间，仿若城市中心的绿岛一般。大厦的建筑立面上设计了许多由按压成立方体的石化燃料驱动的旧汽车图案，象征着汽车技术的革新时代，还演示了循环利用的过程及媒质的使用。

THREE TOWERS

The program is distributed in two different areas: on one side the lower areas and the basement and on the other side, three slender towers based on the concepts of a stack and a mosaic design where most of the program is located.

A big surface is released in the lower areas with the creation on three slanting surfaces, to create a great public space mimicking a green island in the centre of the city. The façades of the towers are designed by the stack of old cars, based on fossil fuel pressed as cubes which act as icons in the era of change of the automobile technology and example of recycling process and use of media.

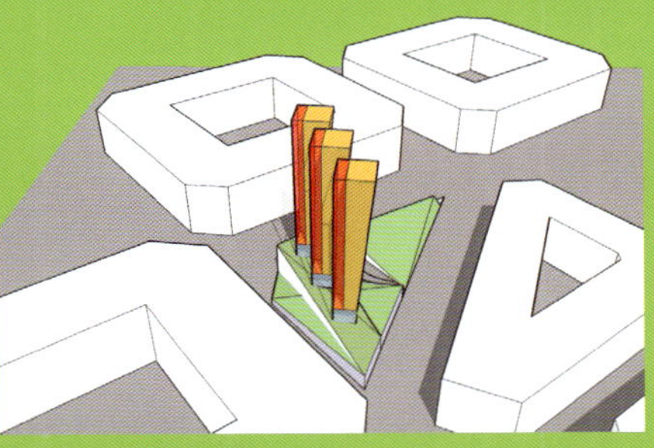

可持续都市交通中心 · 巴塞罗那

Sustainable Urban Transport Centre · Spain

提名奖 · Honourable Mention

110076 (竞标代码)

Estocolmo (团队)

Enric Navarro · Joan Ramon Pastor · David Onieva Alonso · Albert Sabás Serrallonga

公园

该项目主要取意为“城市绿肺”。因此，我们认为有必要掩藏不利于城市可持续发展的大片加油区，而把其改造成为一个公园。其内部空间是公园的延伸空间，使绿化植被穿透圆形的大庭院与白色的墙体交相辉映。

A PARK

The main image of the project is based on the idea to promote a green lung for the city. Thus, we thought it was important to hide the massive refuel area, unsuitable of a sustainable city. The buried building turns into a park. The interior space is understood as an extension of the park, where the green penetrates the big round courtyards and merges with the white walls.

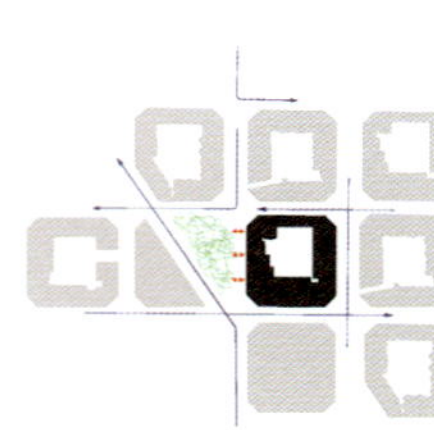

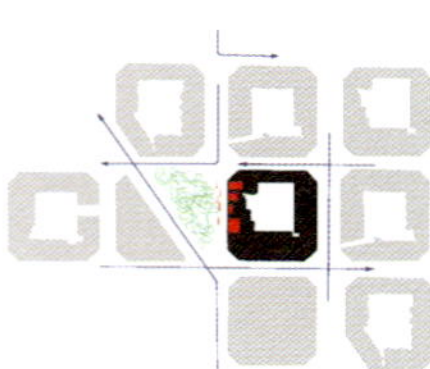

地块位置 SITE PLAN

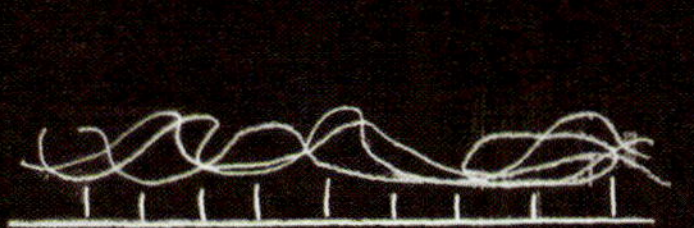

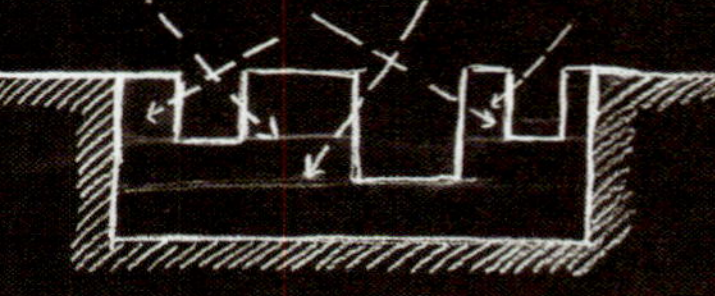

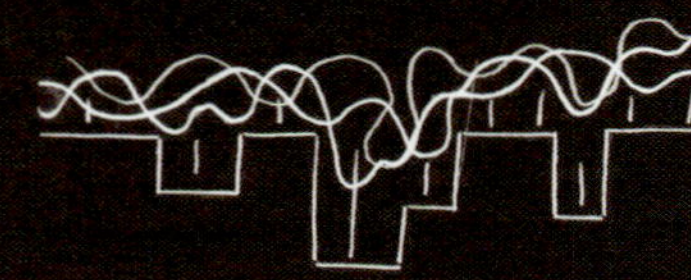

方案概念 IDEA OF THE PROJECT

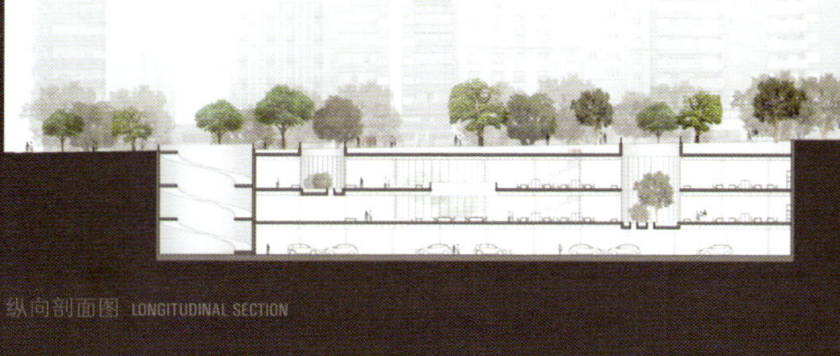

纵向剖面图 LONGITUDINAL SECTION

横向剖面图 CROSS SECTION

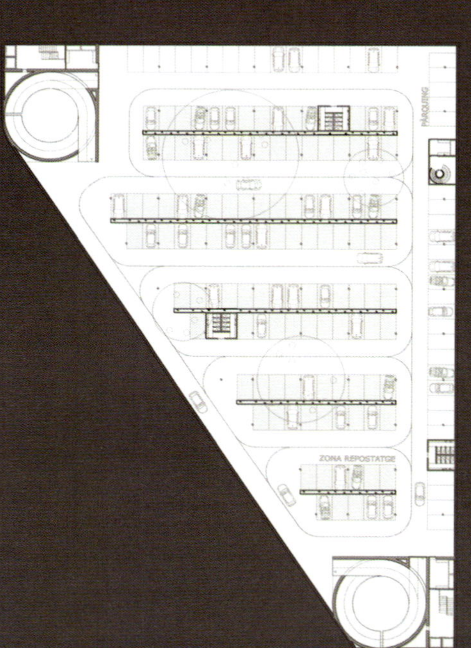

停车场层平面图 PARKING FLOOR PLAN

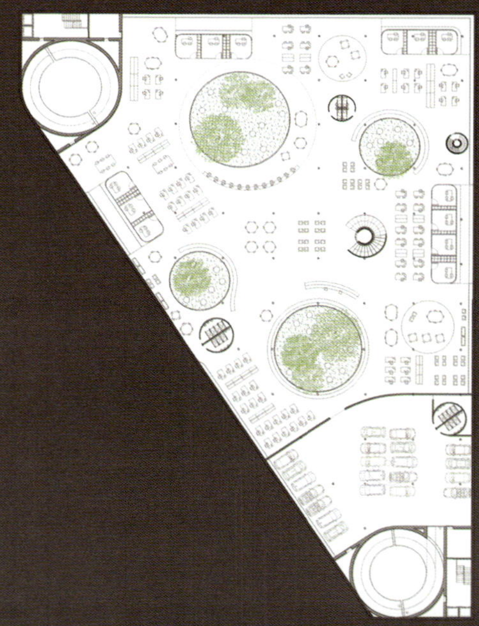

地下一层平面图 BASEMENT FLOOR PLAN 1

地下二层平面图 BASEMENT FLOOR PLAN 2

可持续都市交通中心·巴塞罗那

Sustainable Urban Transport Centre · Spain

提名奖 · Honourable Mention

110152 (竞标代码)

GMT+2 (团队)

Sergio del Barco · Raúl Olivares · Pablo Magán
Guillermo Angulo · Marc Cabarga · María Lozano

能源标志

我们打算建造一所新型节能建筑，使其成为巴塞罗那市内的标志性建筑。信息屏将显示能源浪费的实时数据，使市民及时了解能源消耗的不同数值范围。

ENERGETIC ICON

We propose a new energy-efficient building which functions as an icon inside Barcelona city. Information screens show data of the real energetic waste to make citizens aware of the different consumption scales in real time.

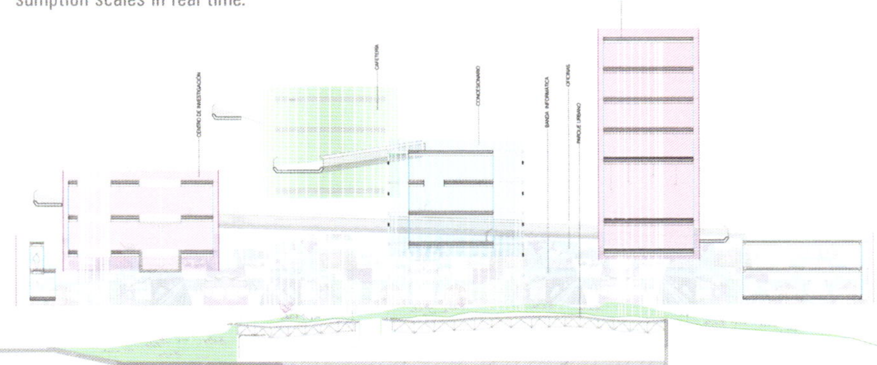

纵向剖面图 LONGITUDINAL SECTION

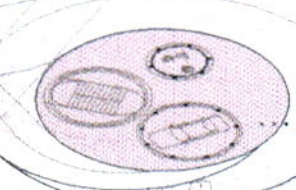

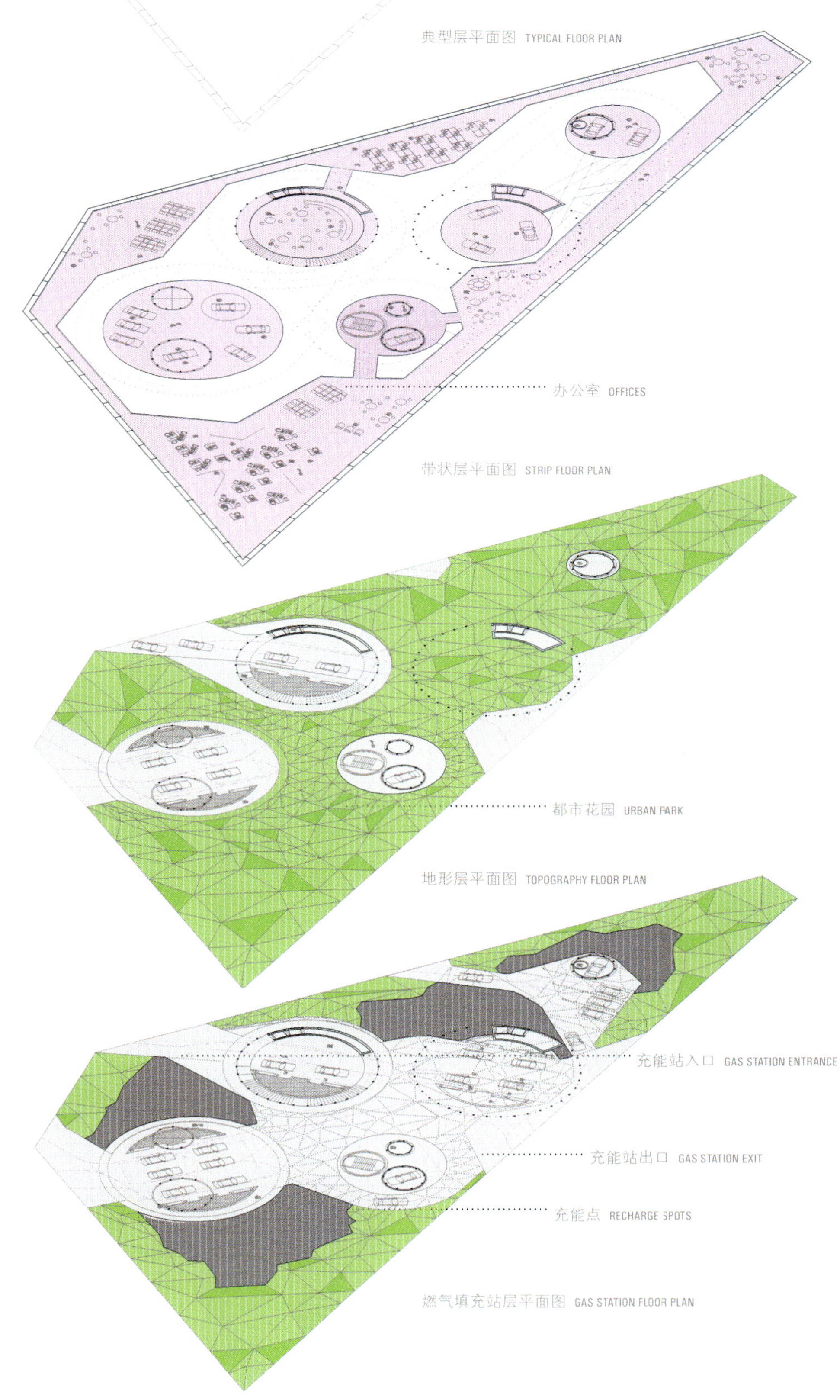

可持续都市交通中心·巴塞罗那

Sustainable Urban Transport Centre · Spain

提名奖 · Honourable Mention

空地

为了实现可持续性交通运输，城市中心拟利用水作为建筑元素，在巴塞罗那的下层土中储备大量水资源，以方便电动车辆的停车和安置之用。建筑意象取自森林中的空地，因此，建筑本身不自成一个单体，而是一个置于背景环境中的结构体。该地段将成为一个环泻湖而建的公共公园，公园里电动车辆、行人和自行车来往穿梭。

A VOID

The urban centre for sustainable transport pretends to use water as an architectural element, generating a big water deposit in the subsoil of Barcelona which can function as a parking and station for electric cars. An architectural icon understood as a void, (a clearing in the forest) which is not another singular and decontextualized building. The site would turn into a public park with a central lagoon where electrical cars, pedestrians and bicycles co-exist.

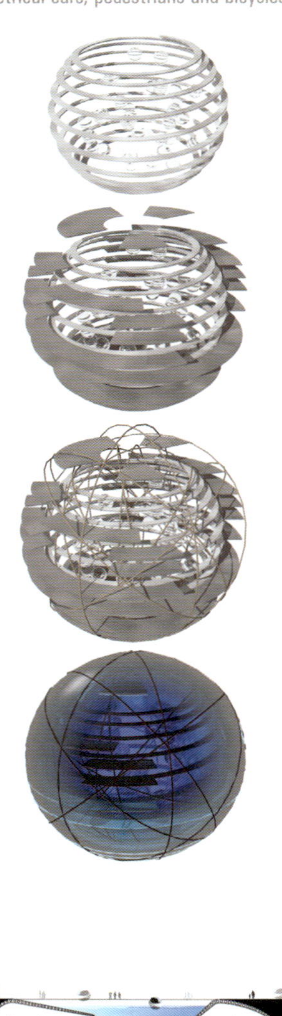

120279 (竞标代码)

Lam (团队)

Lu Lu, Antonio Berrio Gallego, Miguel Garcia Guerra

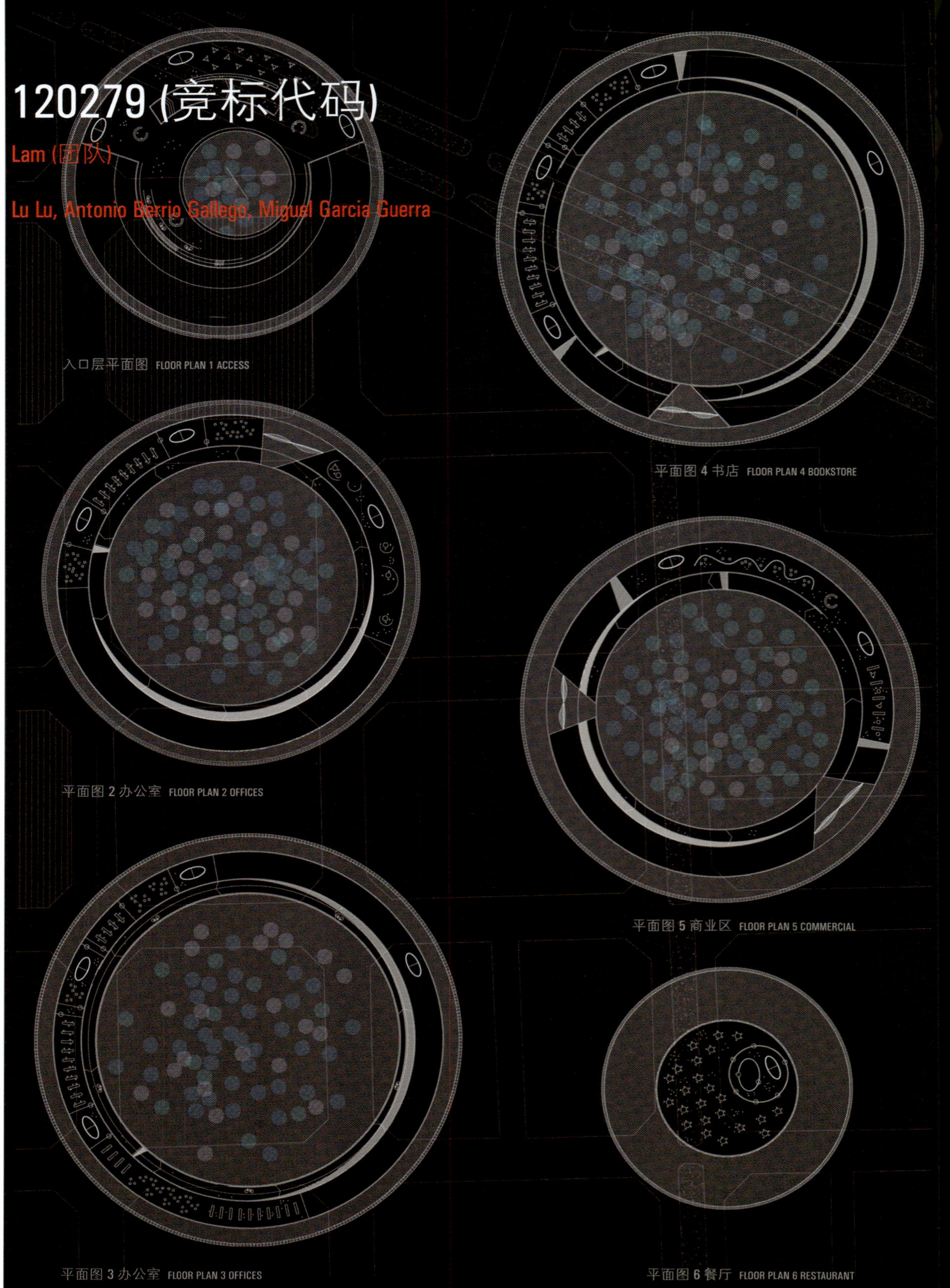

入口层平面图 FLOOR PLAN 1 ACCESS

平面图 4 书店 FLOOR PLAN 4 BOOKSTORE

平面图 2 办公室 FLOOR PLAN 2 OFFICES

平面图 5 商业区 FLOOR PLAN 5 COMMERCIAL

平面图 3 办公室 FLOOR PLAN 3 OFFICES

平面图 6 餐厅 FLOOR PLAN 6 RESTAURANT

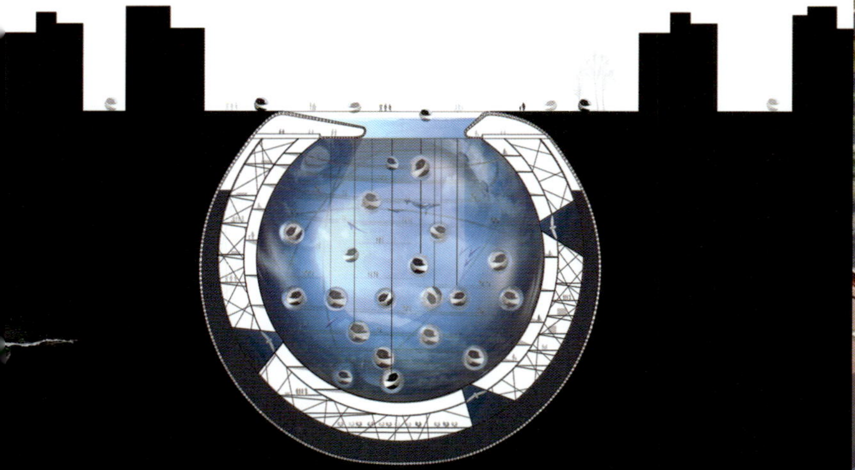

可持续都市交通中心 · 巴塞罗那

Sustainable Urban Transport Centre · Spain

提名奖 · Honourable Mention

110209 (竞标代码)

m+a (团队)

Marcin Giemza · Anna Skrzydlewska

外表生态特征 ECO SKIN FEATURES

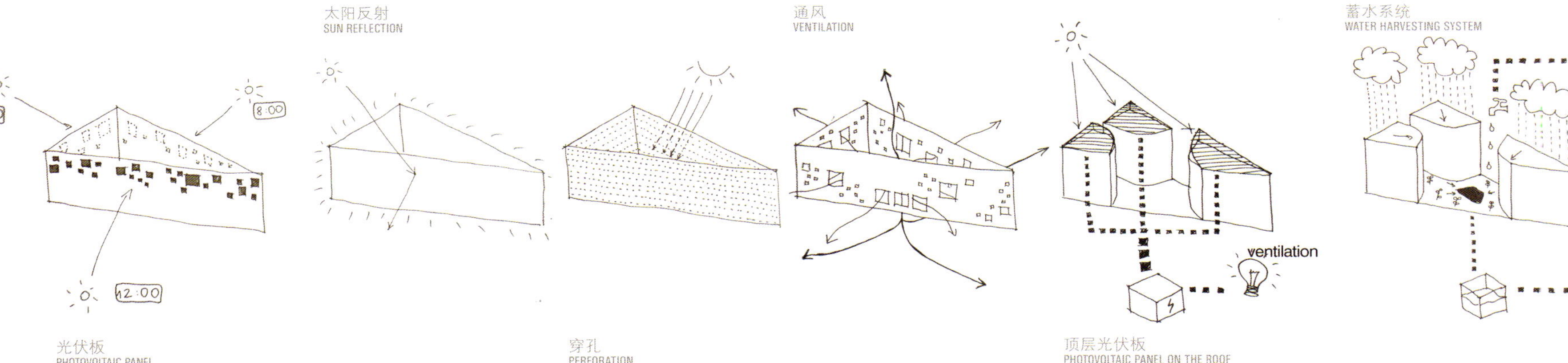

生态表面

该项目由三大部分组成：中心部分包括一个共享空间、一个教育空间和一个交通空间。上方是从中心部分高耸而出的三幢办公楼，还有一个环绕整个建筑结构的生态表面。

ECO-SKIN

The project is based on three main elements: the core, which contains a common space, educational space and the circulation. Above, three office buildings sprout from the core and an eco-skin surrounds the whole volume.

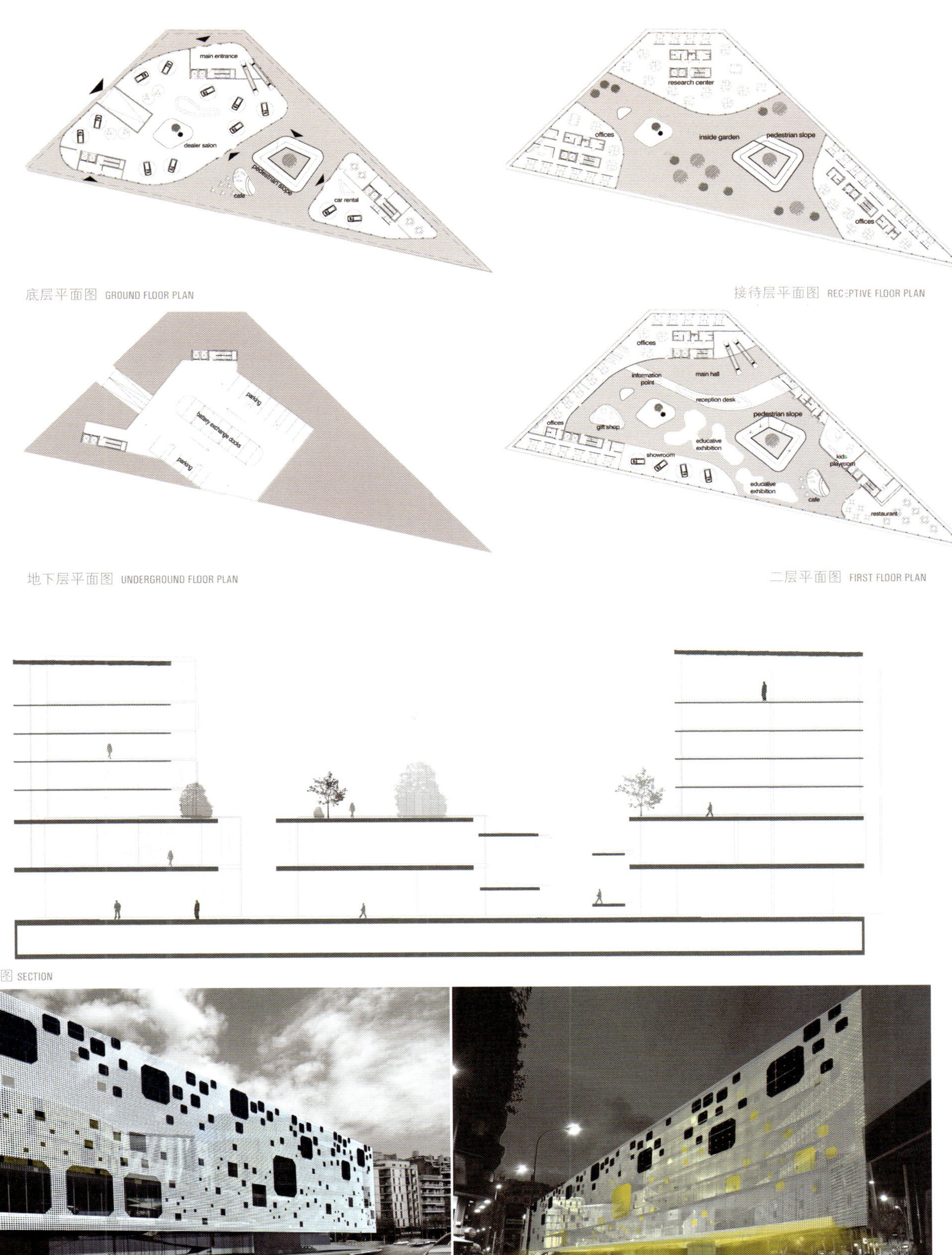

底层平面图 GROUND FLOOR PLAN

接待层平面图 REC:PTIVE FLOOR PLAN

地下层平面图 UNDERGROUND FLOOR PLAN

二层平面图 FIRST FLOOR PLAN

剖面图 SECTION

形状生成 CREATING THE FORM

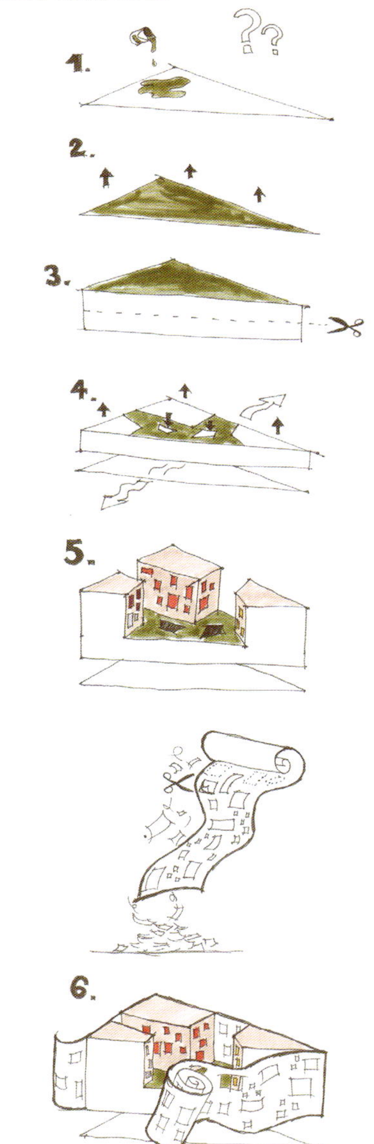

可持续都市交通中心 · 巴塞罗那

Sustainable Urban Transport Centre · Spain

提名奖 · Honourable Mention

lift me up 120313 (竞标代码)

Ortiz + Schoonewolff (团队)

Federico Ortiz · Ricardo Schvonewolff

空中花园

建筑体由一组木质结构支撑悬在空中，上方是一个插入城市建筑物的三角形结构体。内部庭院四方通透，达到了“延伸”的效果。该空间包含一个电动车大型停车场以及一个无明确边界的花园，功能齐全。

A SUSPENDED GARDEN

A series of wooden structures lift the volume of the building as open hands. Above them leans a triangular volume which silently inserts itself inside the urban fabric. Thus the permeability of the inner courtyard is granted, so as to recover one of the main virtues of The Eixample (Catalan for "extension"). This space welcomes a big area to fill up the electric cars and a garden where its limits are blur to integrate the different types of circulation.

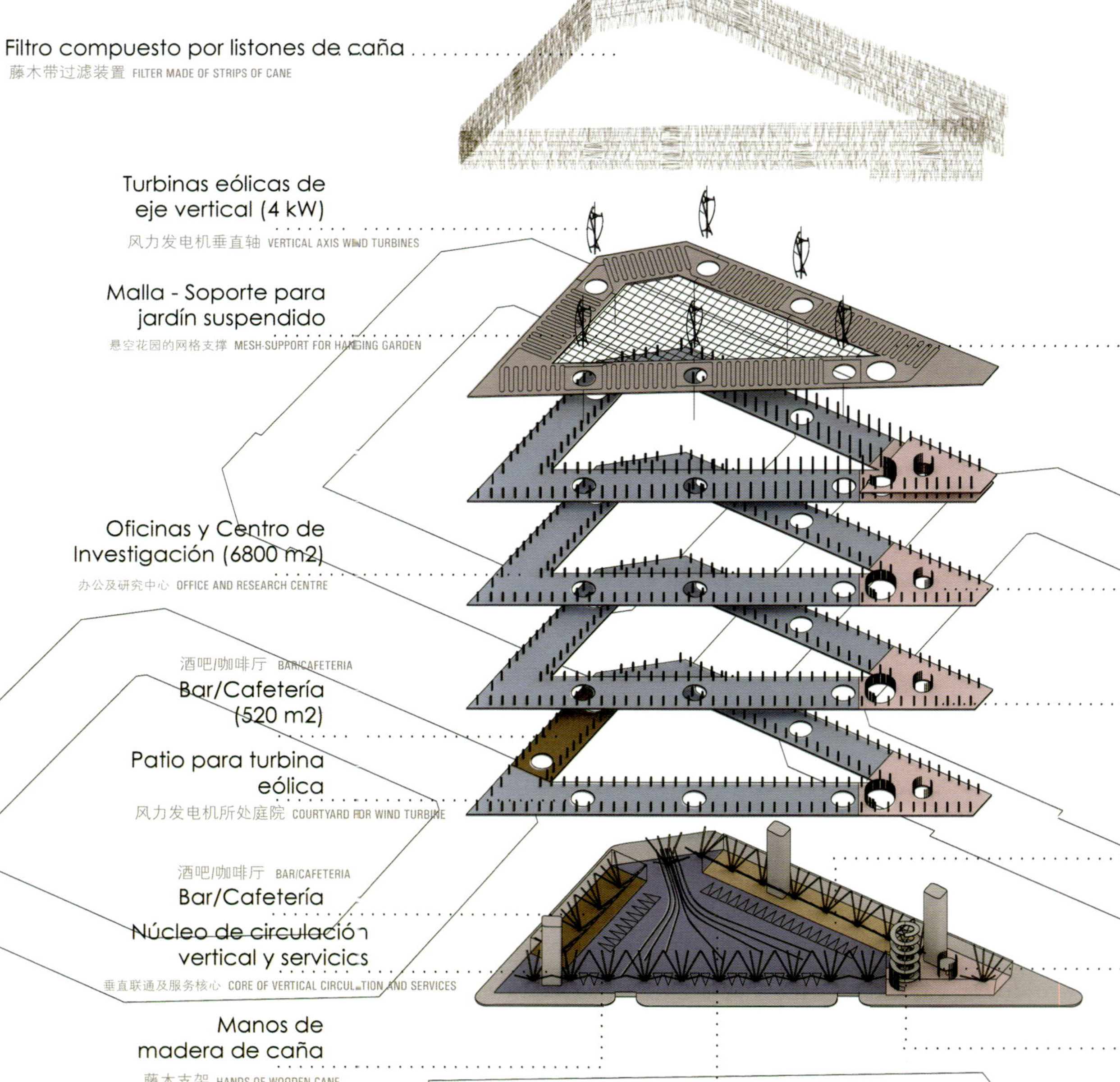

可持续都市交通中心·巴塞罗那

Sustainable Urban Transport Centre · Spain

提名奖 · Honourable Mention

110037 (竞标代码)

Studio X (团队)

Terry Allen Driggs, Jr.

表现纯洁性

建筑表面采用聚氟乙烯打造，这是一种轻型耐用的塑料，呈透明状，可在上面印刷底纹和广告。聚氟乙烯也可用作太阳能电池板进行发电。整个结构完全由轻型牢固的碳纤维打造而成，代表了建筑未来的发展趋势。

DESIGNED TO REPRESENT PURITY

The skin would be constructed of ETFE, a kind of lightweight and durable plastic that can be transparent or printed on for shading or advertising. ETFE can also be used as solar panels to produce energy. The structure would be made entirely of carbon fiber, a lightweight but extremely strong material to signify the buildings steps towards tomorrow's building.

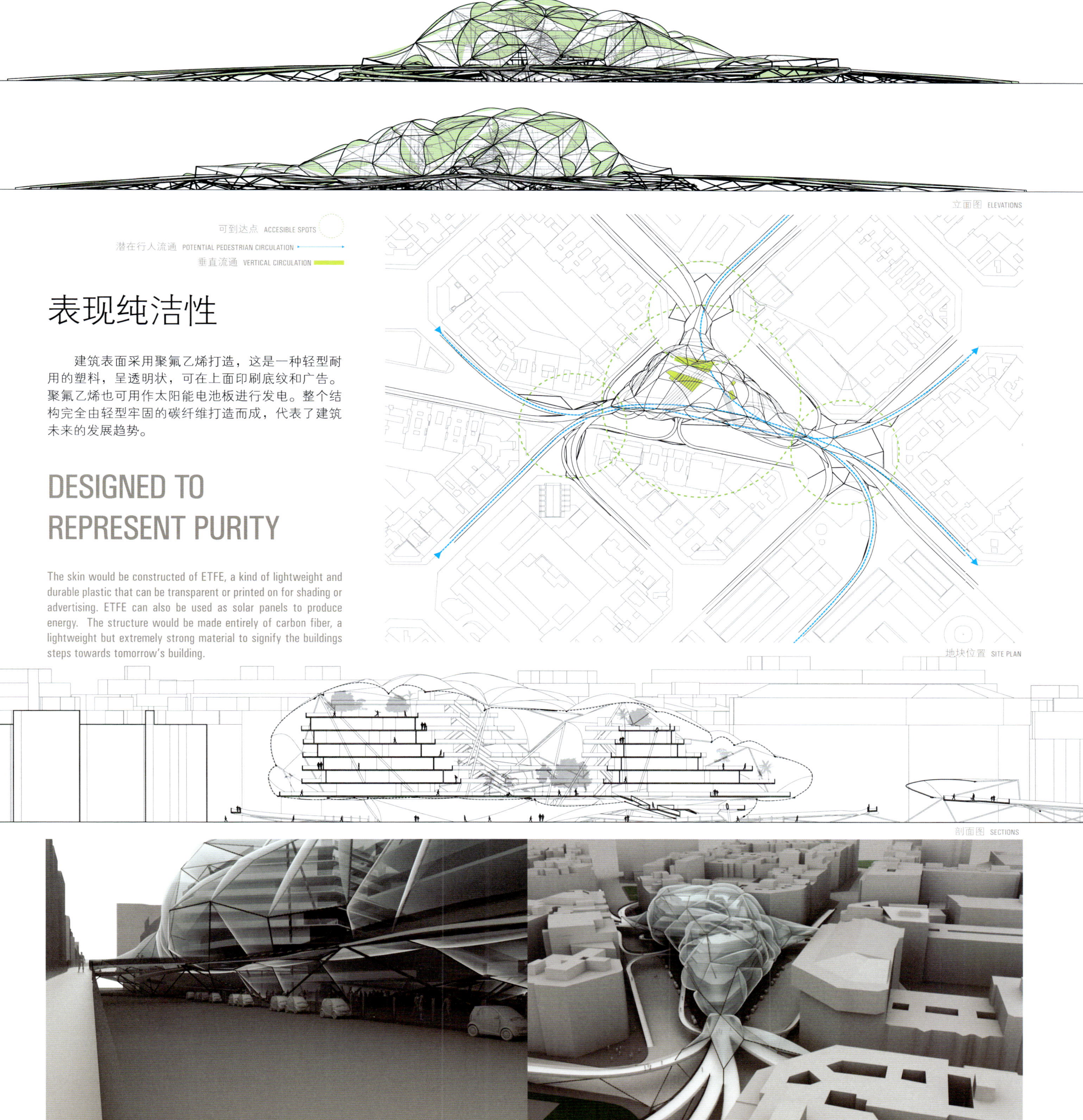

可持续都市交通中心 · 巴塞罗那

Sustainable Urban Transport Centre · Spain

提名奖 · Honourable Mention

110037 (竞标代码)

Unoencinco (团队)

Alejo Lopez · Noelia Maldonado · Laura Abate · Miguel Urruty · Maximo Triolo

弥补 LINE UP | 释放 LIBERATION | 地下占用 OCCUPATION OF THE SUBSOIL | 自然光 NATURAL LIGHT | 加油 FILL UP WITH FUEL | 底层入口 GROUND FLOOR ACCESS | 研究所 RESEARCH OFFICE | 研究所露台 RESEARCH TERRACE

宣言

自从第一次城市文化运动以来，城市就成为了人类社会、文化、政治以及技术改革的主要发生地。该项目旨在打造一个大型城市空间，宣告向合理利用能源理念的转变。

A MANIFESTO

Since the generation of the urban culture in the cities-polis of the first civilizations, the urban agoras have been the main scenarios of the social, cultural, political and technological transformations of the human race. Thus, this project proposes the creation of a big urban space which can function as a manifesto sup this change towards a rational use of the energy.

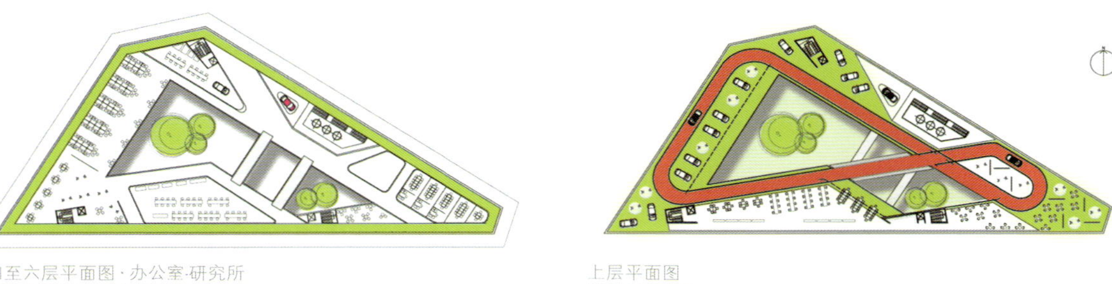

四至六层平面图 · 办公室-研究所
FLOOR PLANS 3-5 · OFFICES-INVESTIGATION

上层平面图
UPPER FLOOR PLAN

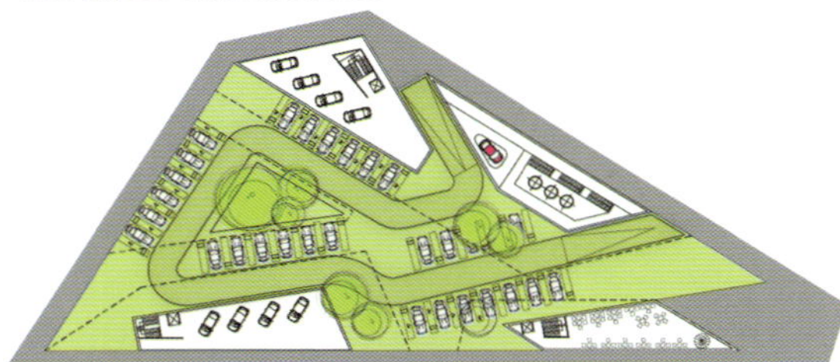

地下层平面图 · 生态加油
UNDERGROUND FLOOR PLAN · ECOLOGICAL FILLING UP

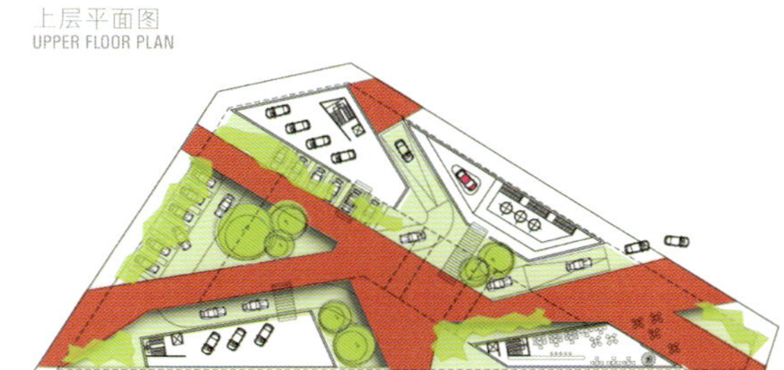

底层平面图 · 公共空间
GROUND FLOOR PLAN · PUBLIC SPACE

可持续都市交通中心 · 巴塞罗那

Sustainable Urban Transport Centre · Spain

入围 · Finalist

110038 (竞标代码)

Anxinal (团队)

Iván Valero Fernández · Laura Domènech López

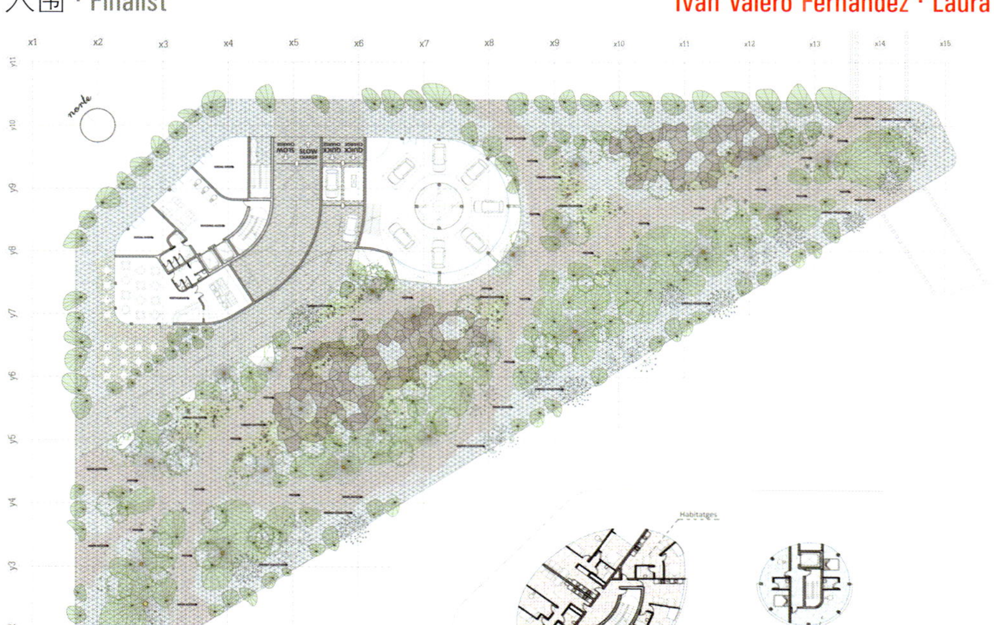

底层平面图 GROUND FLOOR PLAN

十一层平面图 FLOOR PLAN 10

附加值

大部分的太阳能采集所需面积被释放出来，从而可以使建筑更集中发展其高度并创造新的公共空间(汽车零售展示，办公室，公寓及商铺)。高度与方位的结合使建筑拥有一个太阳能采集外墙面，用以满足车辆所需加载的大部分能量。

VALUE ADDED

Most of the plot area is liberated and the area is thus concentrated in its height (two towers), generating livable new public spaces thanks to the mixture of programs of the towers (concessionaire-exhibition, offices, housing and commercial). The height, combined with the orientation allows building a solar façade which can generate most of the energy necessary to recharge the cars.

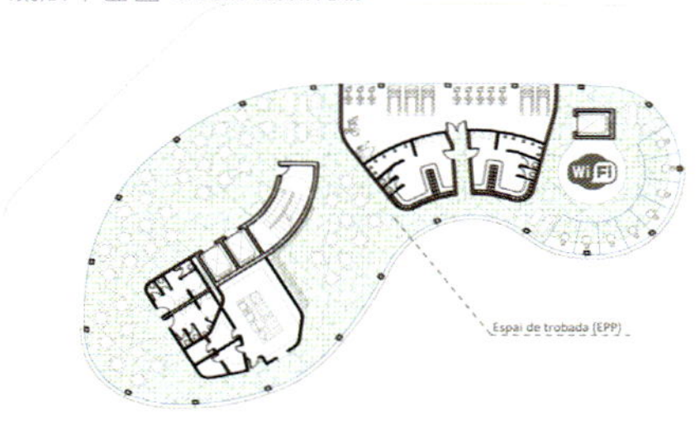

七层平面图 FLOOR PLAN 6

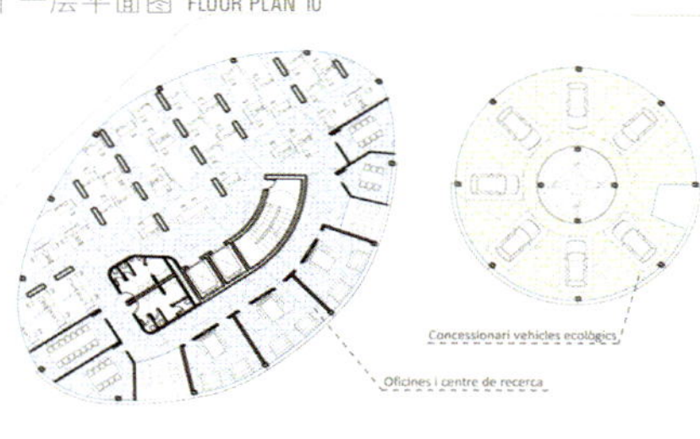

五层平面图 FLOOR PLAN 4

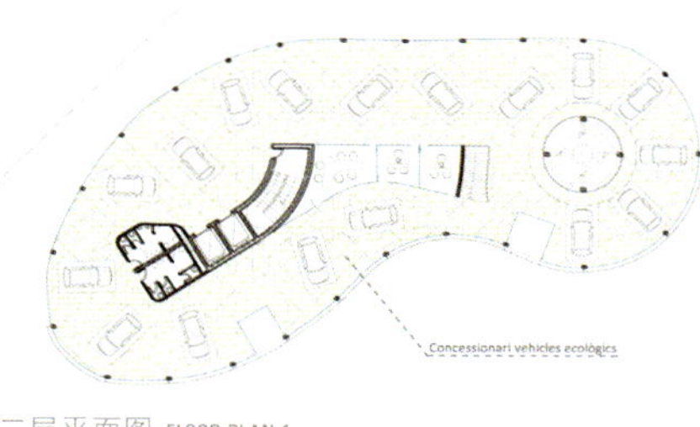

二层平面图 FLOOR PLAN 1

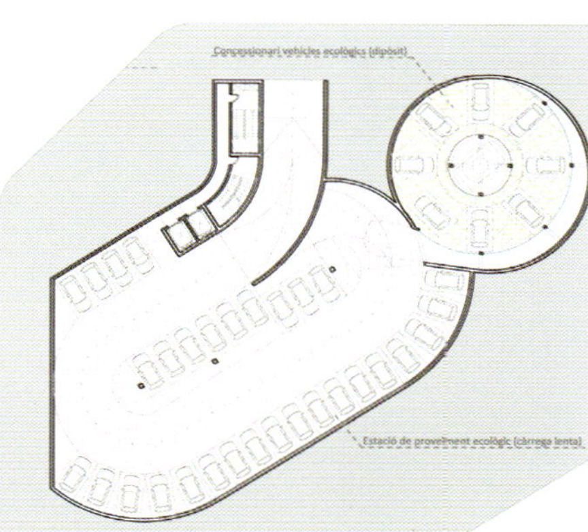

地下层平面图 UNDERGROUND FLOOR PLAN

新贝加莫中心

New Headquarters of The Province of Bergamo · Italy

竞标 · competition
新贝加莫中心
New Headquarters of The Province of Bergamo

竞标类型 · competition type
资格限定一轮制匿名竞标
restricted procedure in a single anonymous stage with prequalification of participants

项目地点 · site area
贝加莫省 · 隆巴底地区 · 意大利 Province of Bergamo · Lombardy region · Italy

主办方 · promoter
贝尔加墨省 Provincia di Bergamo
城市规划基建部 Settore Pianificazione territoriale, Urbanistica e Grandi Infrastrutture

日程安排 · schedule
招标 · Announcement 01.2009
评审结果 · Jury's results 05.2009

评审团 · jury

Prof. Alberto Castoldi (教授)
Arch. Giuseppe Epinati (建筑师)
Arch. Aurelio Galfetti (建筑师)
Arch. Giuseppe Gambirasio (建筑师)
Arch. Giuseppe Marinoni (建筑师)
Ing. Andrea Moltrasio (工程师)
Arch. Vito Sonzogni (建筑师)

获奖者 · awards

一等奖 · first prize
Arata Isozaki & Associates, Arata Isozaki Associati (建筑师事务所)
设计 design: 矶崎新 Arata Isozaki · Andrea Maffei

设计团队 design team · Mauro Mazzali · Alessandra De Stefani · Carlotta Maranesi · Takeshi Miura Higaki Seisuke – Arata Isozaki Associati srl, Milano, Arata Isozaki & Associates Co. Ltd., Tokyo
结构 structure: ing. Sandro Favero · Favero & Milan Ingegneria spa
规划 plans: ing. Manuele Petranelli, Spring srl
顾问 consultants: Alessandro Rocca · Catherine Mosbach · Bruno Gritti

入围 · finalist
Josep Llinas i Carmona (建筑师)

入围 · finalist
Gonçalo Byrne + Pedro Sousa (建筑师事务所)
现场建筑师 local architects · Gabriele Cappellato · ARCODE · Marisa Macchietto · Katia Accossato · Luigi Trentin
合作 (c) Alicia Lazzaroni · Laura Hernandez · Manuela Novo · Carla Lousada · Davide Duarte
景观 landscape: Proap · João Nunes

入围 · finalist
Office For Metropolitan Architecture (建筑师事务所)
合伙人经理 partners-in-charge · Rem Koolhaas
Reinier de Graaf

联合负责人 associate-in-charge · Richard Hollington III
团队 team: Fred Awty · Katrin Betschinger · Philippe Braun · Ippolito Pestellini Lawrence Siu
结构 structure: Werner Sobek Stuttgart GmbH & Co. KG
预算 cost consultant: Studio Tre Architetti Associati
现场顾问 local consultant: Matteo Poli

入围 · finalist
OBR Open Building Research (建筑师事务所)
合伙人经理 partners-in-charge: Paolo Brescia · Tommaso Principi

合作 (c) OBR Open Building Research · Paolo Brescia · Tommaso Principi · Carlotta de Bevilacqua Buro Happold · Martha Schwartz Partners · Aubry & Guiguet Programmation · TRM Engineering S.r.l.
团队 team: Yellow Office · Studio Tre Architetti Associati · Pietro Bagnoli · Paolo Brescia Tommaso Principi · Michele Renzini · Margherita Menardo · Fabiano Cocozza · Fabio Valido Chiara Farinea · Giulia D'Ettorre · Paula Vier · Jeannette Sordi · Marta Rolando · Izabela Sobjerai · Matteo Guidi · Laura Pesaro · Laura Pessoni · Silvia Zanolini · Paola Monaco

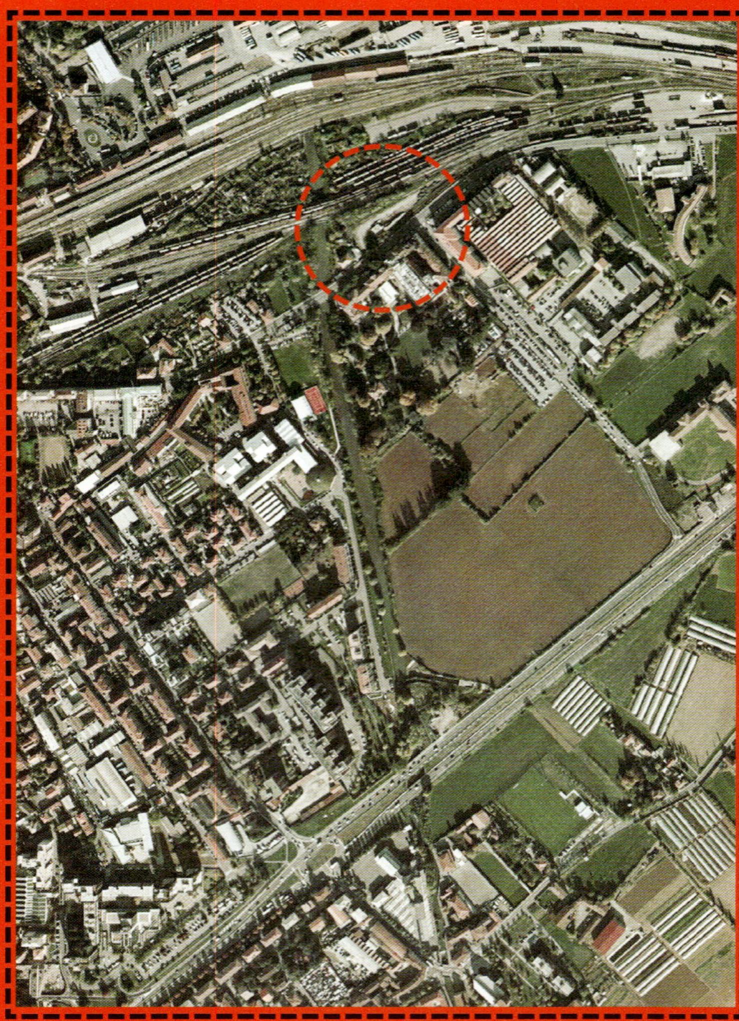

新贝加莫中心是一个城景和风景相互改造的工程，对位于北边的火车站和位于南边的Mugazzone Way进行重新改造，使这两大区域彼此间相得益彰。该中心占地0.9平方千米，周边地区和全省均在现有的城市风貌基础上作了战略性的地位转换。新建筑将新贝加莫中心的两部分连为一体，结构上严密规整，象征意义上对应相通。

The new center of Bergamo is a project of urban and landscape transformation and requalification of the areas between the train station to the north and the Mugazzone Way to the south. Almost 90 hectares in a strategic exchange position within the existing city, the surrounding territory and the entire Province. **The new building is an integral connection, geometrical and symbolical**, of the two parts of the new center of Bergamo.

新贝加莫中心

New Headquarters of The Province of Bergamo · Italy

一等奖 · First Prize

Arata Isozaki & Associates · Arata Isozaki Associati (建筑师事务所)

Arata Isozaki · Andrea Maffei (建筑师)

横向摩天大楼

我们可能不相信离市中心很近的地方居然有一幢仅有90米高的建筑物，但这种建筑物对贝加莫这样的城市来说再合适不过了。在贝加莫，新修建筑物的楼高更低，与美轮美奂的Upper Town以及层峦叠嶂的山麓交相辉映。确实，这个新工程在设计时不仅仅参照了当代贝加莫城，还考虑到了新建筑应突破铁轨的概念，开启城市发展的新篇章。合约规定，该建筑物不得高于88米，但要在竞争限制条件下表现出所要求的水平横向的建筑立面。按照这个方案，该工程不是一幢垂直耸立的独栋高楼，不会呈现出厚重感，将与整个城市的风貌风格协调一致。

HORIZONTAL SKYSCRAPER

We do not believe that a building nearly 90 meters high in this specific location, not far from downtown, is the most appropriate for a city as Bergamo, where the height of the new buildings is much lower, and should especially enjoy the extraordinary presence of the Upper Town and the hills. It is true that the reference of the new project cannot only be the current Bergamo, especially since the new building was also entrusted with the task to launch a new phase of urban development beyond the railway track. The building does not rise up to 88m, as permitted by contract, but develops the required horizontal surfaces within the limits of the competition. This solution releases the project from a heavy impact without adopting the kind of monolithic vertical tower, which would be intrusive on the skyline of the city.

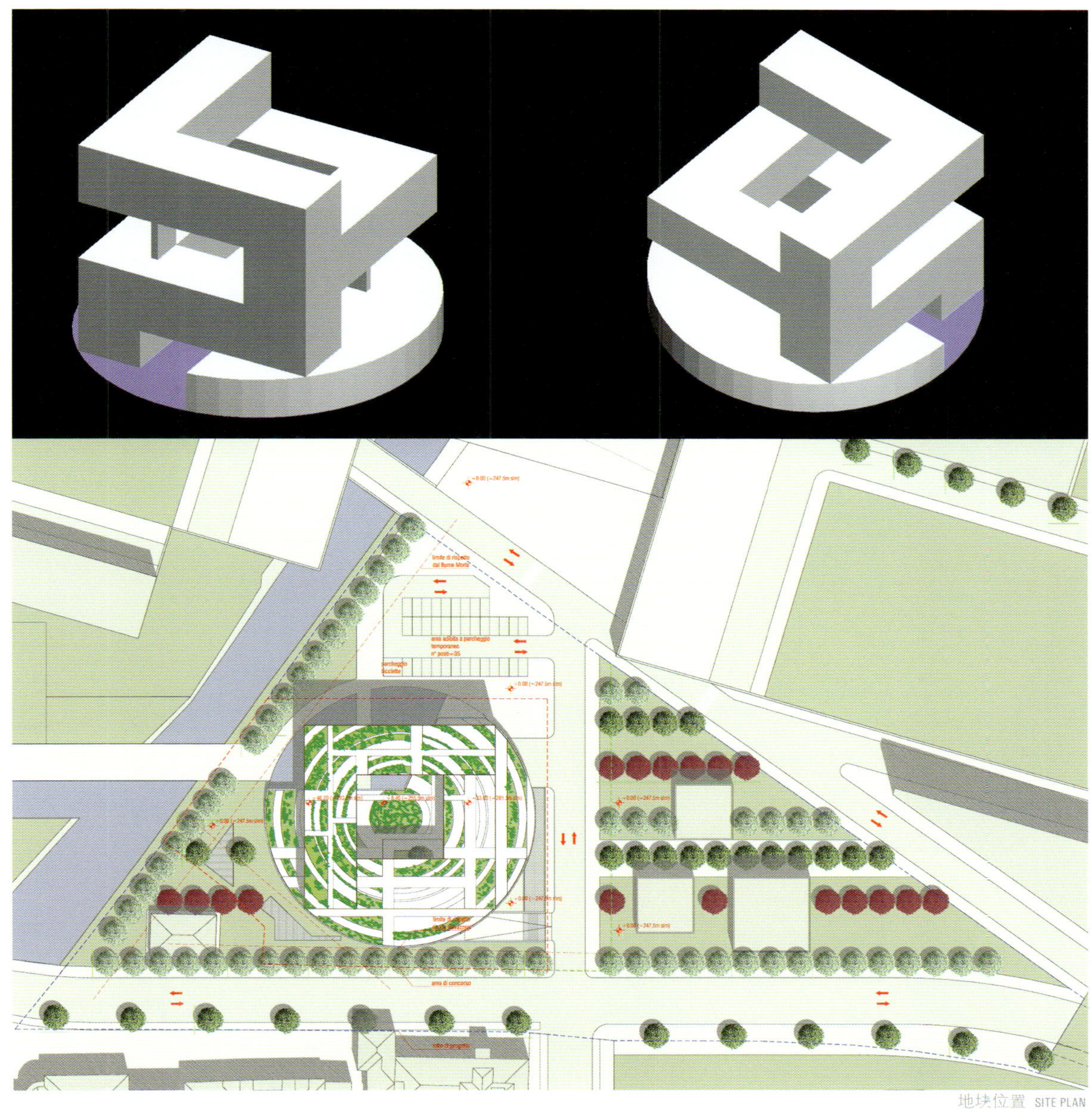

地块位置 SITE PLAN

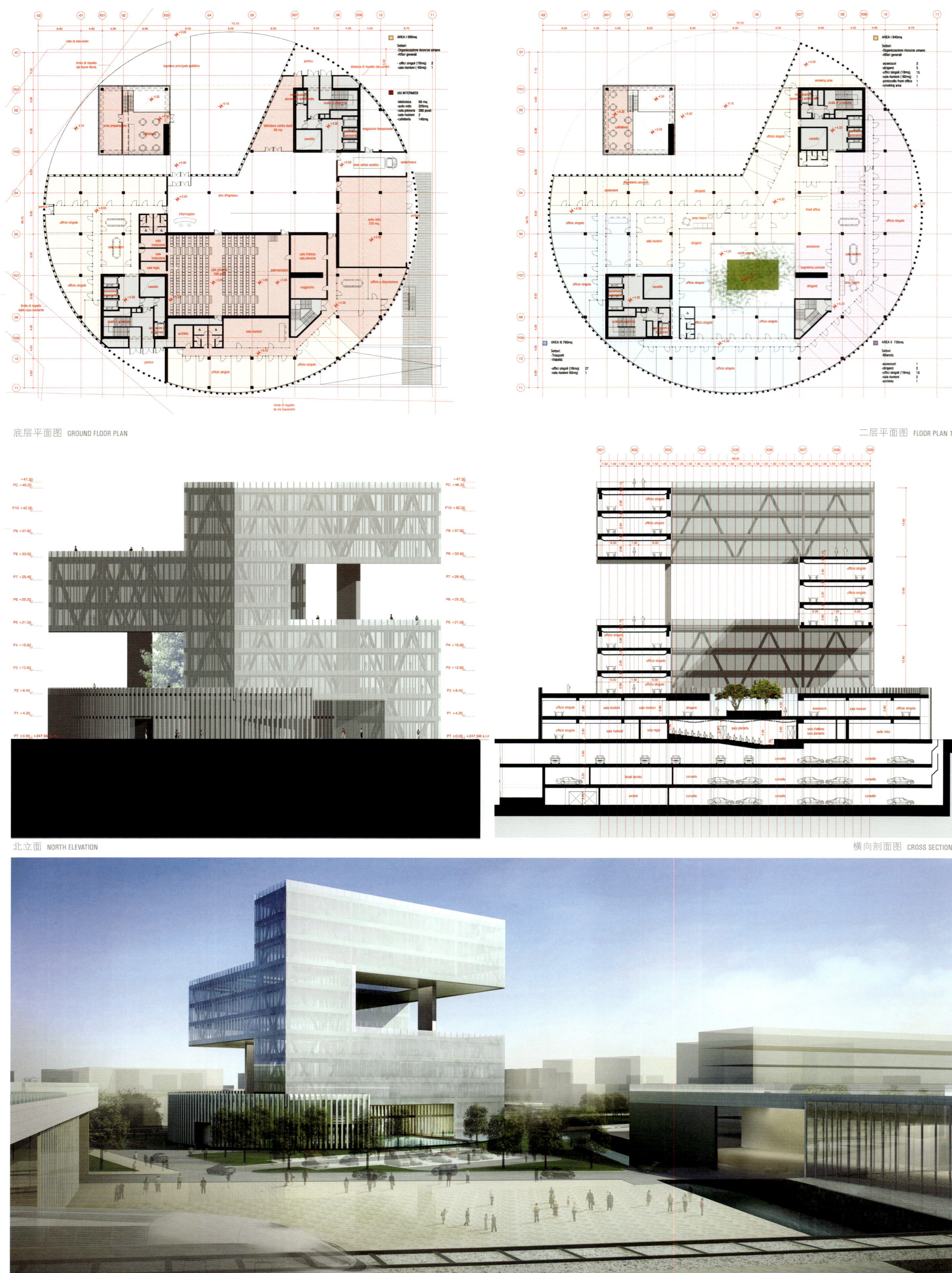
底层平面图 GROUND FLOOR PLAN
二层平面图 FLOOR PLAN 1
北立面 NORTH ELEVATION
横向剖面图 CROSS SECTION

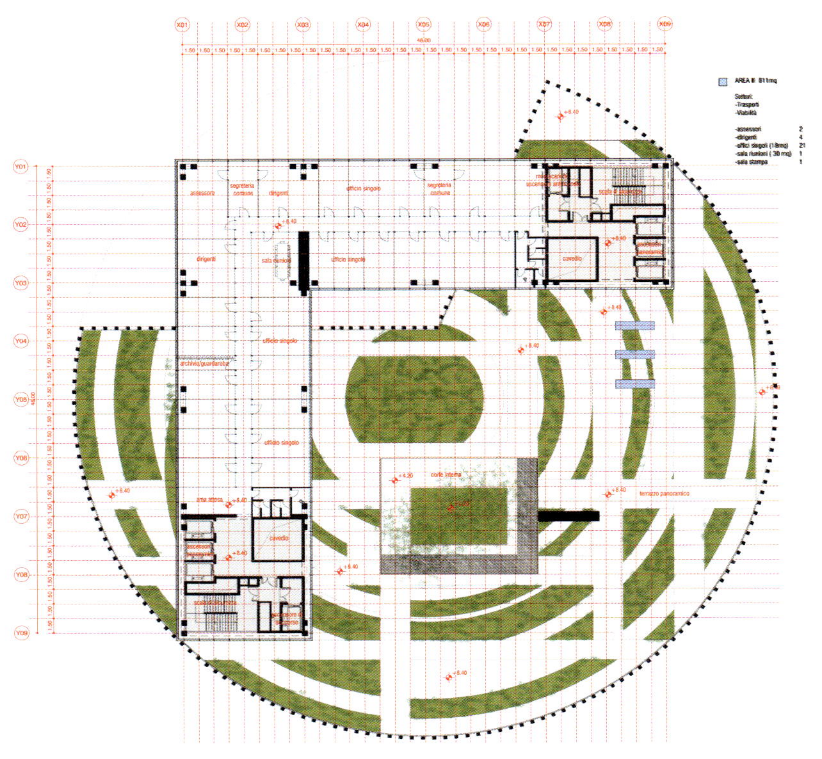

三层平面图 FLOOR PLAN 2

六层平面图 FLOOR PLAN 5

新贝加莫中心

New Headquarters of The Province of Bergamo · Italy

入围 · Finalist

17935 (竞标代码)

Josep Llinas i Carmona (建筑师)

两种模式

我们提出了城市发展的两种模式。一种是结合现今城市局限性制订的紧密围绕铁路基础设施建设的模式，另一种模式是建筑体按不同的组合方式排列在绿色空间中。按这两种模式打造出的“矛盾”区域在视觉效果上看不出棱角分明的几何结构。

TWO TYPES

We propose two types of urban developments; a compact limit to the railway infrastructure in contact with the current limit of the city and another in which the building blocks are arranged in a pattern of green spaces defined by different alignments. These two alignments create an area of "friction" with non-orthogonal geometries.

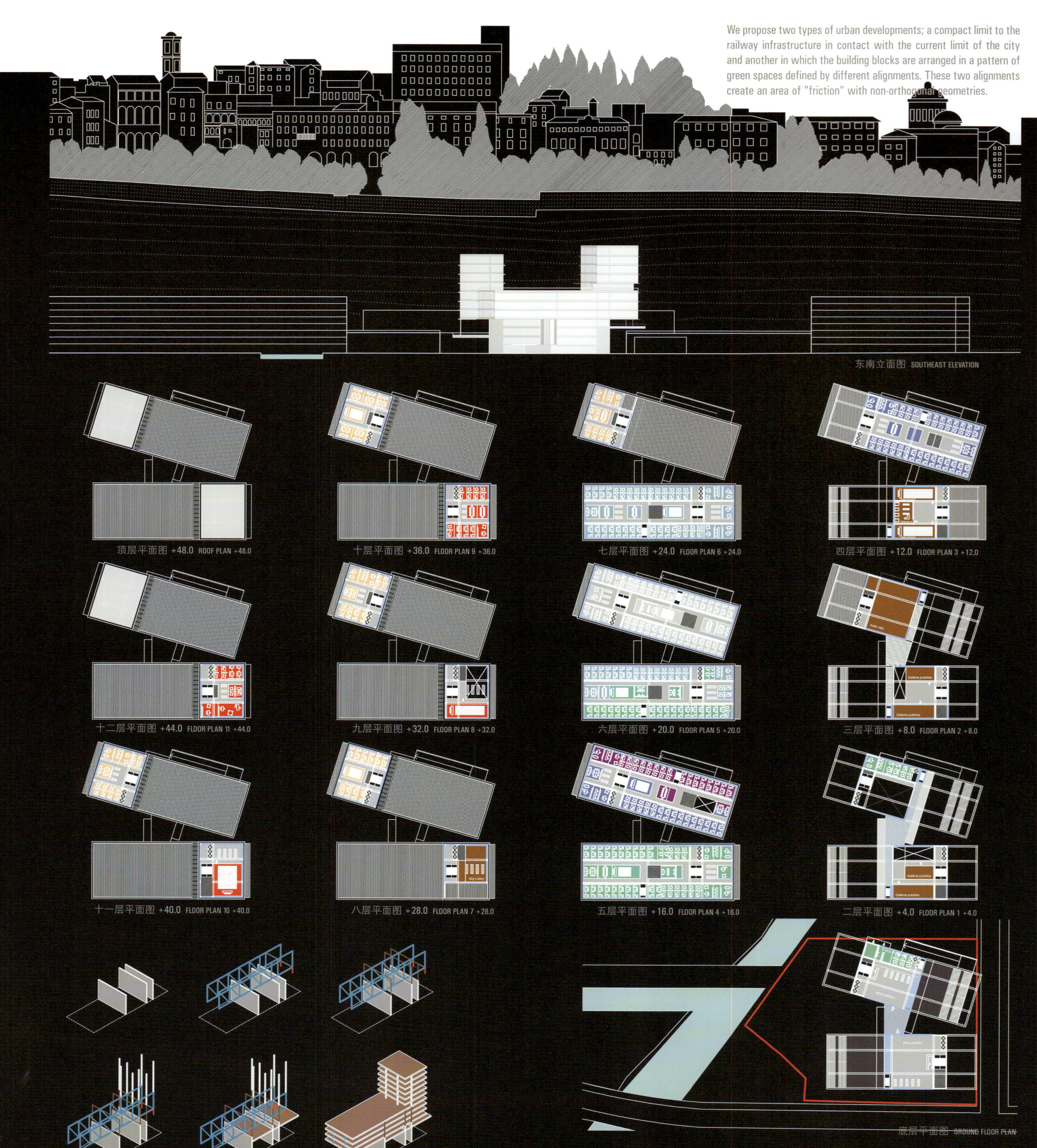

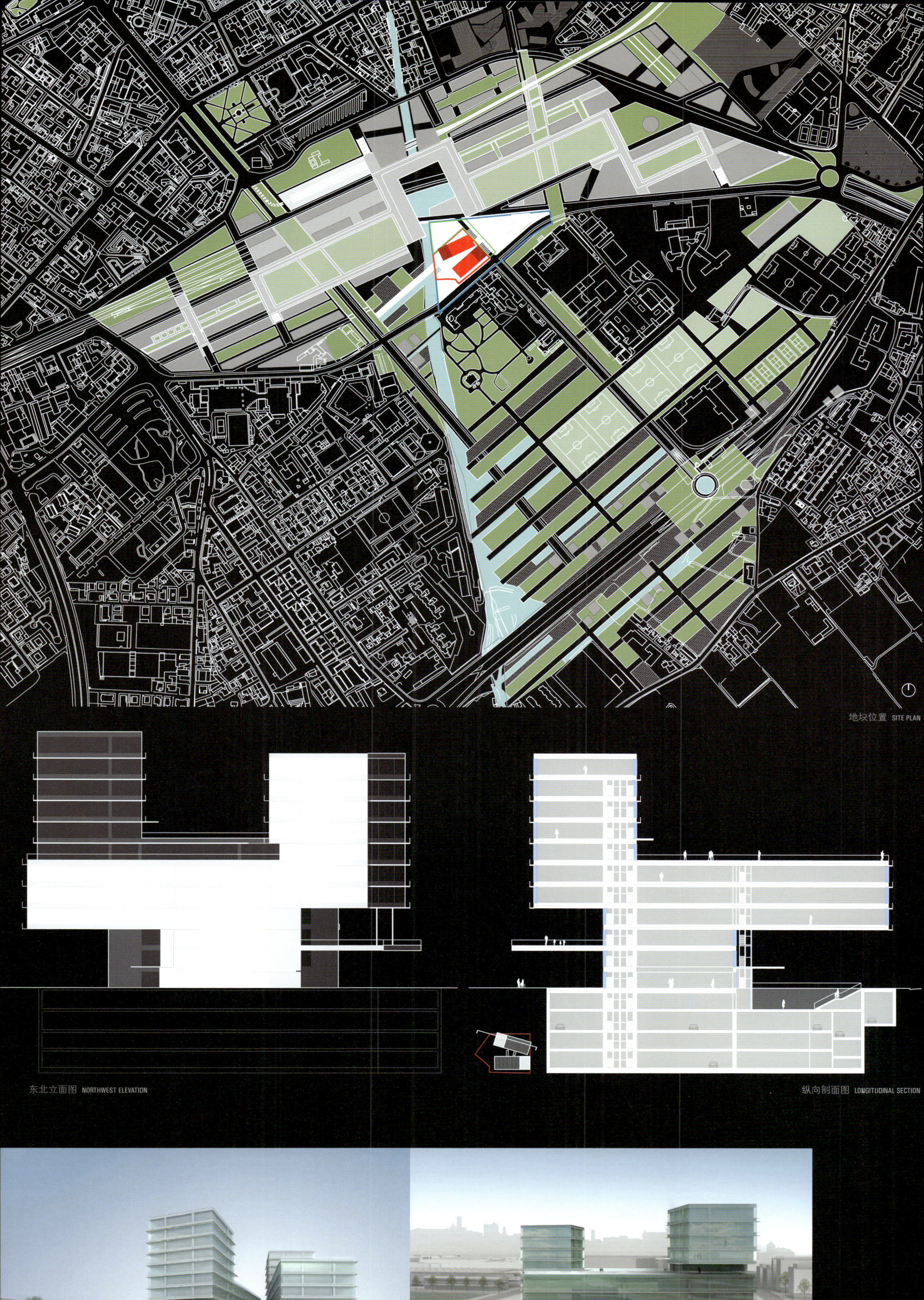

地块位置 SITE PLAN

东北立面图 NORTHWEST ELEVATION

纵向剖面图 LONGITUDINAL SECTION

新贝加莫中心

New Headquarters of The Province of Bergamo · Italy

Gonçalo Byrne + Pedro Sousa (建筑师)

入围 · Finalist

视觉孔隙

从历史角度来分析，筹划在贝加莫新建一个总部时必须充分考虑两大元素：Upper Town分界的城墙与Medieval Towers。这两大元素在设计理念上相互融合，这样，高楼看上去没有任何偏向性定位，风格统一，既能保证多个不同的视角，又能有效构建促成物景合一的通道。

VISUAL POROSITY

The historical analysis has suggested and underlined the presence of two elements that were essential in considering plans for creating the new headquarters in Bergamo. These two elements include the wall, which sets the limit of the Upper Town, and the Medieval Towers. These two elements combine in a conceptual way, creating a tower without any preferential directions that always guarantees different perspectives and permeable passages towards the landscape.

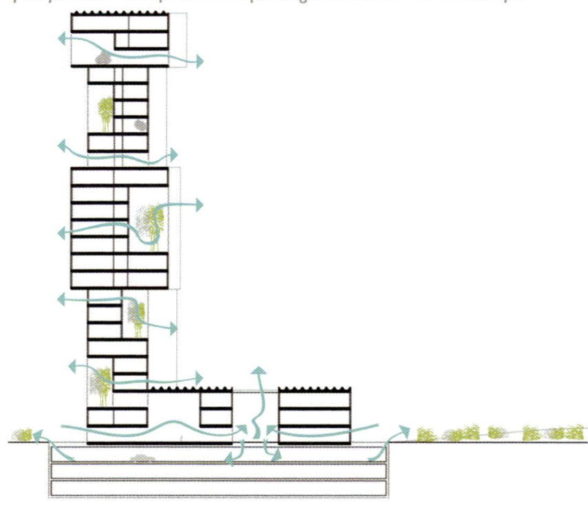

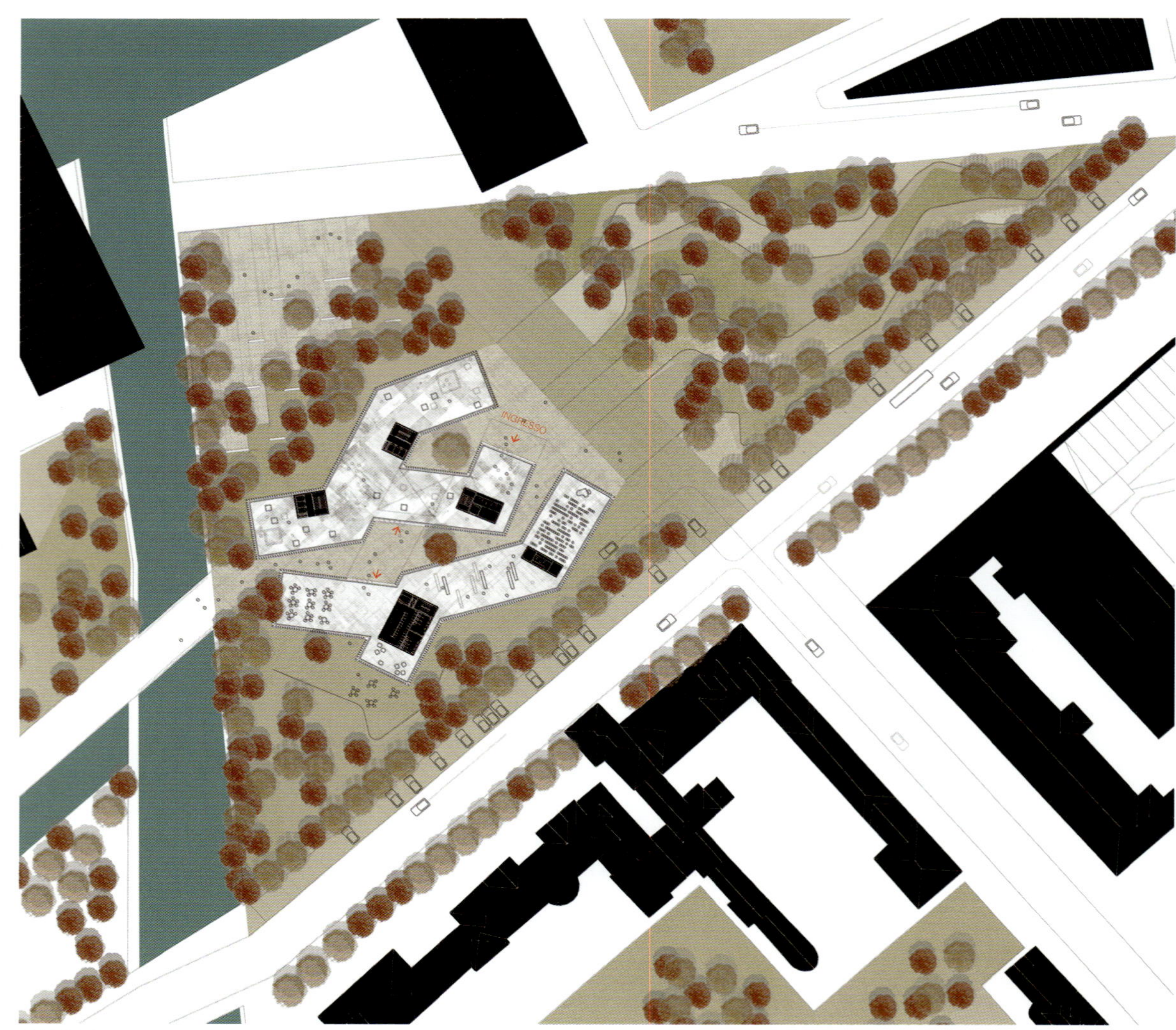

地块位置 SITE PLAN

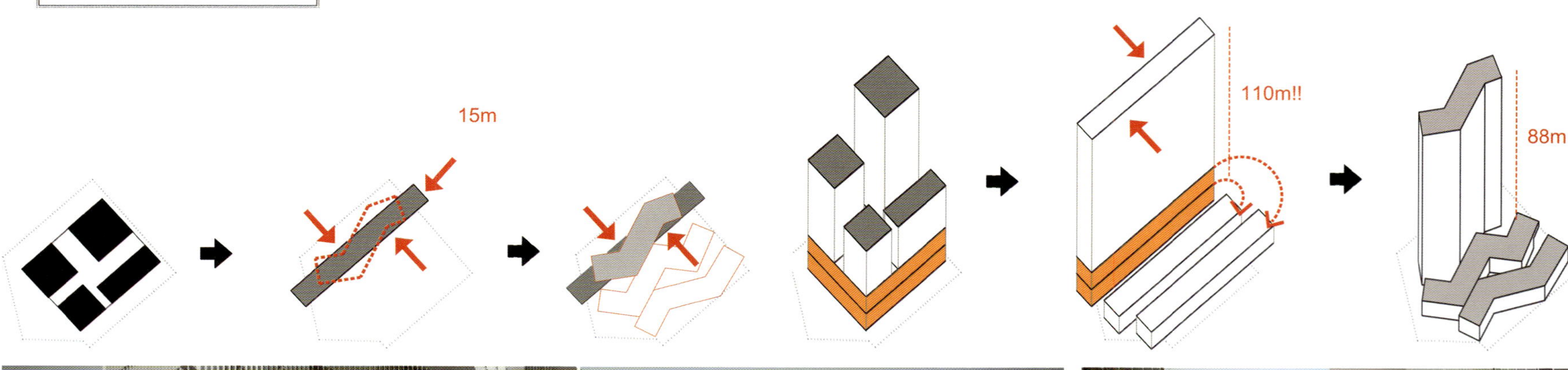

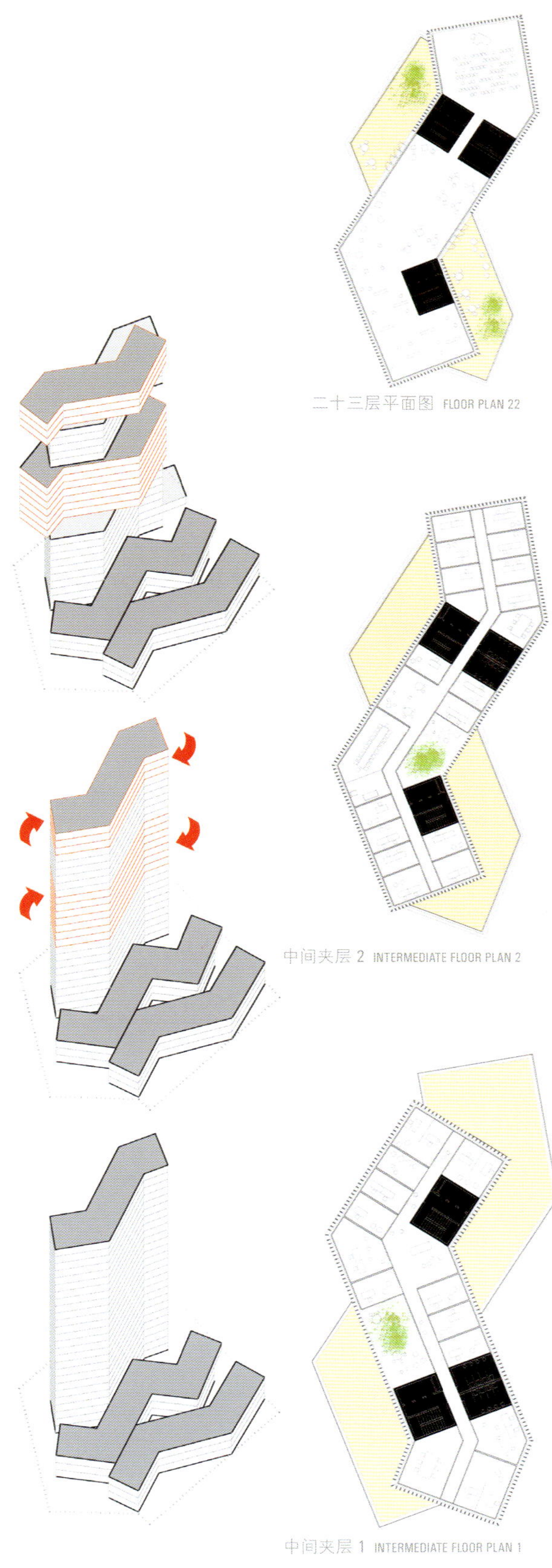

二十三层平面图 FLOOR PLAN 22

中间夹层 2 INTERMEDIATE FLOOR PLAN 2

中间夹层 1 INTERMEDIATE FLOOR PLAN 1

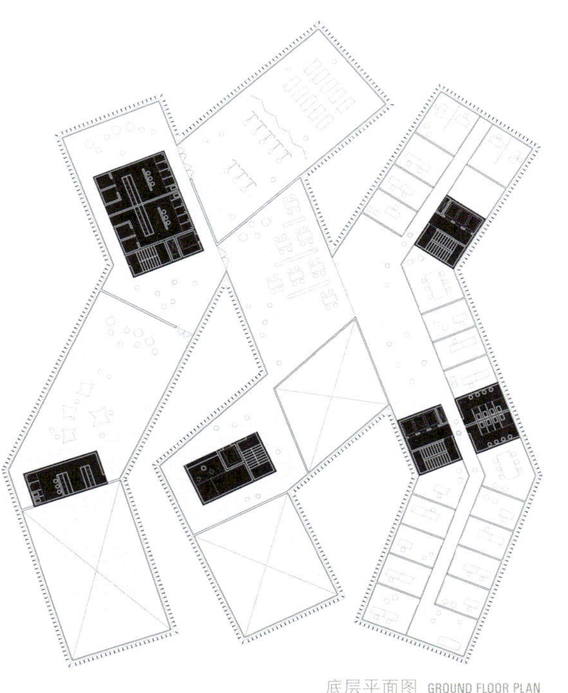

底层平面图 GROUND FLOOR PLAN

新贝加莫中心

New Headquarters of The Province of Bergamo · Italy

入围 · Finalist

35379 (竞标代码)

Office for Metropolitan Architecture OMA (建筑师)

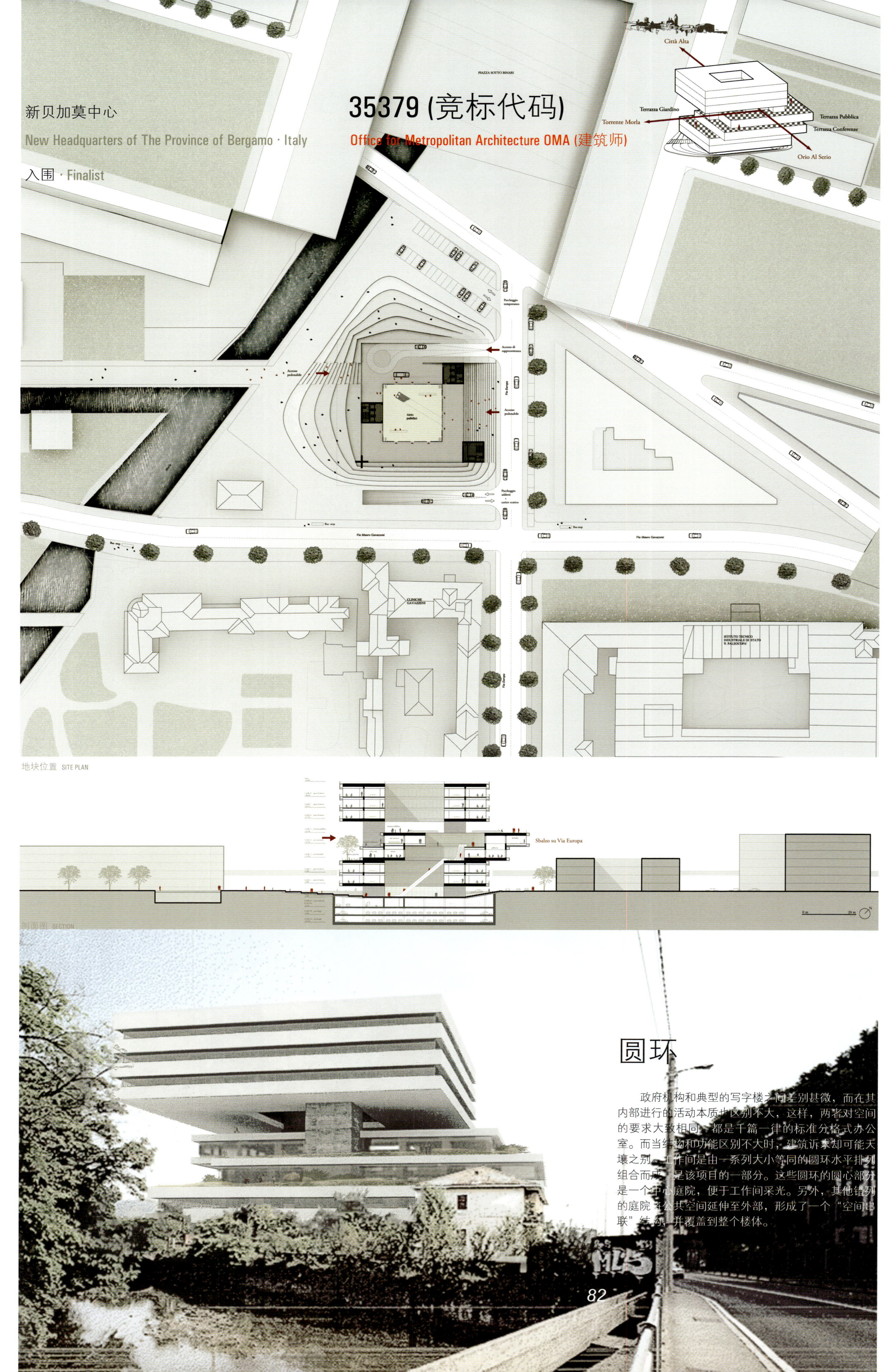

地块位置 SITE PLAN

剖面图 SECTION

圆环

政府机构和典型的写字楼之间差别甚微，而在其内部进行的活动本质也区别不大，这样，两者对空间的要求大致相同，都是千篇一律的标准分格式办公室。而当结构和功能区别不大时，建筑诉求却可能天壤之别。工作间是由一系列大小等同的圆环水平排列组合而成，是该项目的一部分。这些圆环的圆心部分是一个中心庭院，便于工作间采光。另外，其他错列的庭院将公共空间延伸至外部，形成了一个“空间串联”结构，并覆盖到整个楼体。

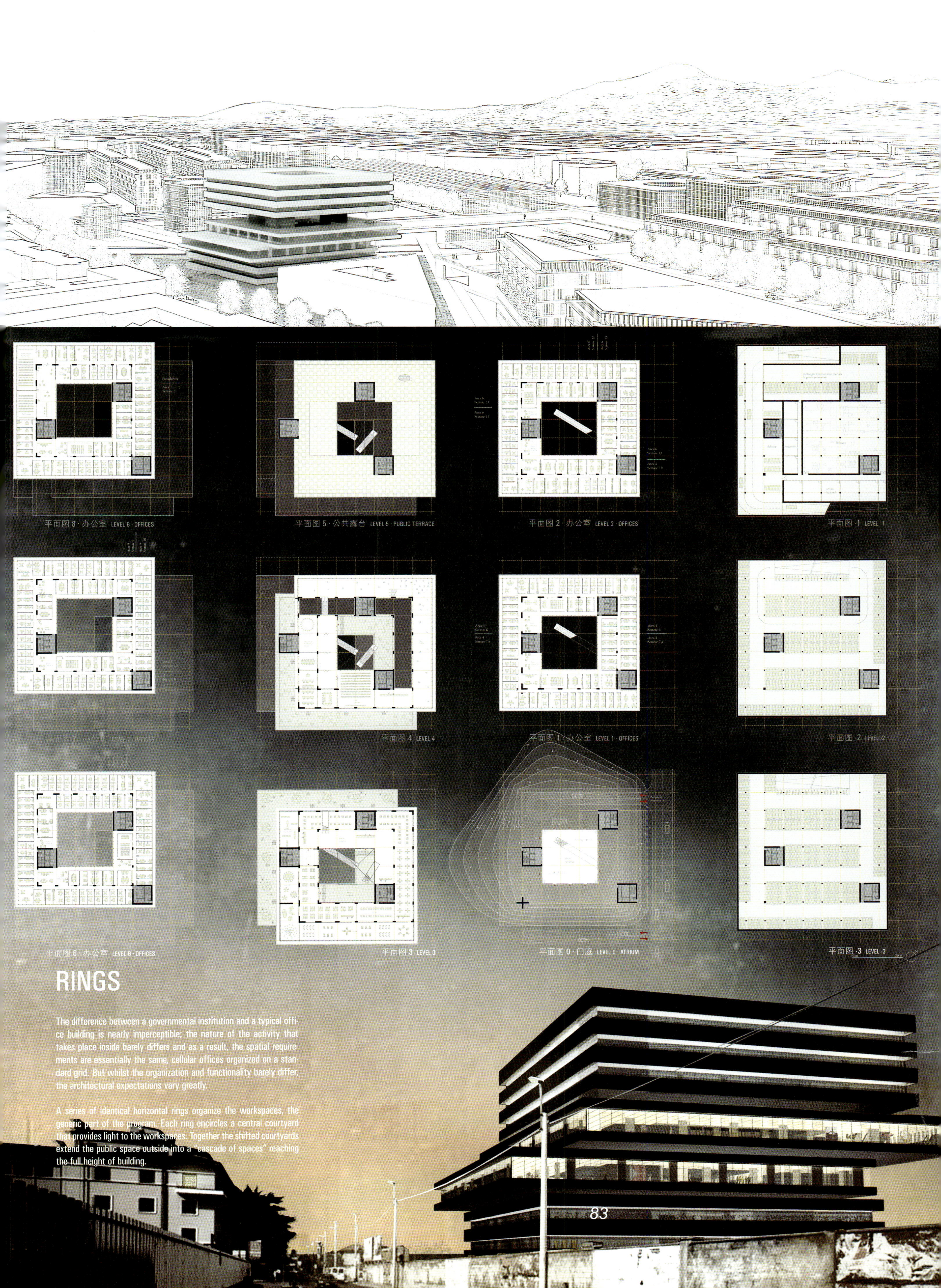

RINGS

The difference between a governmental institution and a typical office building is nearly imperceptible; the nature of the activity that takes place inside barely differs and as a result, the spatial requirements are essentially the same, cellular offices organized on a standard grid. But whilst the organization and functionality barely differ, the architectural expectations vary greatly.

A series of identical horizontal rings organize the workspaces, the generic part of the program. Each ring encircles a central courtyard that provides light to the workspaces. Together the shifted courtyards extend the public space outside into a "cascade of spaces" reaching the full height of building.

新贝加莫中心

New Headquarters of The Province of Bergamo · Italy

入围 · Finalist

57705 (竞标代码)

OBR Open Building Research (建筑师事务所)

Paolo Brescia · Tommaso Principi (建筑师)

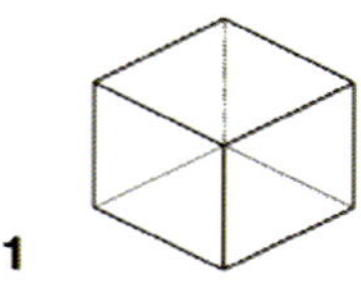
1

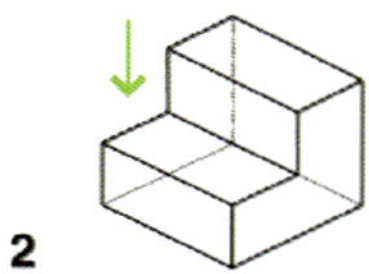
2

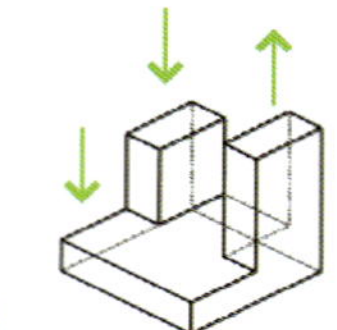
3

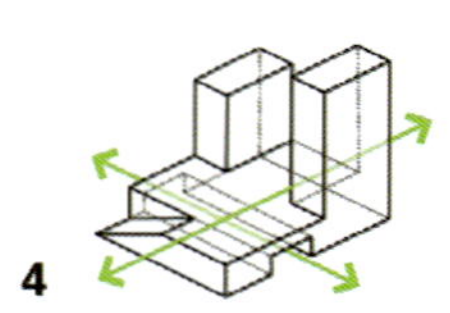
4

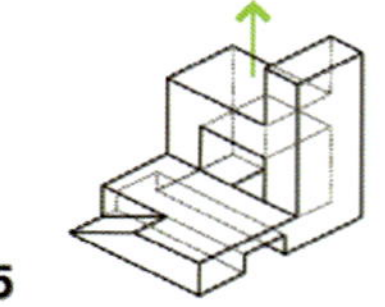
5

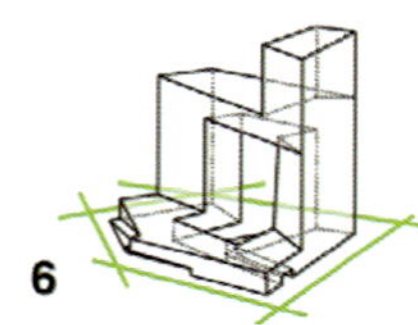
6

整合

新建筑阐释了“群”和“孔隙”的理念，改善了密集型城市的新中心和周围景色之间的关联。在此背景下，新建的总部成为了一幢地标性建筑，是城市革新进程中在该方面的标新立异之举。

INTEGRATION

The new building interprets the ideas of “cluster” and “porosity”, improving the relationships of the new centre with the compact city and the surrounding landscape. In this context the new headquarter becomes a landmark, a new founding act of the urban renewal of this part of the city.

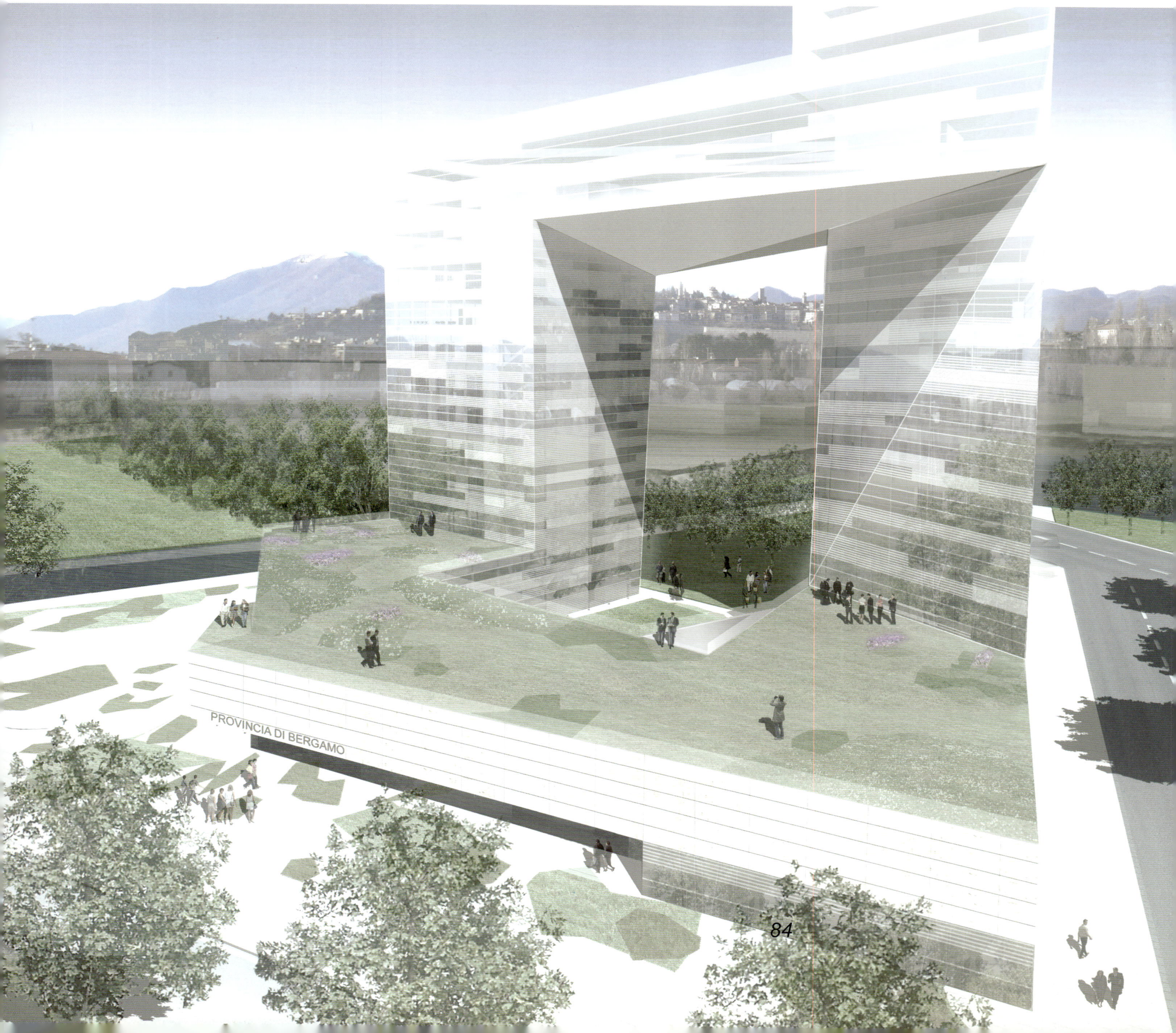

二层平面图 LEVEL 1

四至十层平面图 LEVEL 3-9

底层平面图 LEVEL 0

三层平面图 LEVEL 2

市府新行政大楼 · 塔林

New administrative building of the city Government · Estonia

竞标 · competition
塔林市府新行政大楼
New administrative building of the Tallinn City Government

竞标类型 · competition type
两轮制概念公开竞标
open, two-stage design competition

项目地点 · site area
塔林 · 爱沙尼亚 Tallinn · Estonia

主办方 · promoter
城市规划局及爱沙尼亚建筑师联盟
City Planning Department and the Union of Estonian Architects

日程安排 · schedule
招标 · Announcement 08.2008
评审结果 · Jury's results 06.2009

评审团 · jury

主席 President:
Taavi Aas
Endrik Mänd · 首席建筑师 Architect in Chief, 塔林市政府 Tallinn City Government
Viljar Meister · 行政主任 Administrative Director, 塔林市政府 Tallinn City Government
Janis Dripe · 首席建筑师 Architect in Chief, 里加市政府 Riga City Government
Tarald Lundevall · 建筑师 Architect, 建筑师事务所 Snohetta, 挪威 Norway
Peter Wilson · 建筑师 Architect, 建筑师事务所 BOLLES+WILSON GmbH&Co.KG
Martin Aunin · 建筑师 Architect, 爱沙尼亚建筑师联盟 Union of Estonian Architects
Tiit Trummal · 建筑师 Architect, 爱沙尼亚建筑师联盟 Union of Estonian Architects
Kalle Komissarov · 建筑师 Architect, 爱沙尼亚建筑师联盟 Union of Estonian Architects
back-up member: Andres Levald · 建筑师 Architect, 爱沙尼亚建筑师联盟 Union of Estonian Architects
秘书 secretary: Pille Epner · Project manager, 爱沙尼亚建筑师联盟 Union of Estonian Architects

获奖者 · awards

一等奖 · first prize
BIG - Bjarke Ingels Group (建筑师事务所)
合伙人经理 partner-in-charge : Bjarke Ingels
主案负责人 project leader : Jakob Lange

团队 team: Ondrej Janku · Hanna Johansson · Daniel Sundlin · Harry Wei · Alex Cozma · Jin-Kyung Park · Ken Aoki · Benjamin Engelhardt · Maxime Enrico · Frederik Lyng · Joao Albuquerque
Aet Ader · Steve Huang

二等奖 · second prize
Alver Architects (建筑师事务所)

Indrek Rünkla · Sven Koppel · Ulla Saar · Tarmo Laht · Andres Alver

三等奖 · third prize
3+1 Architects (建筑师事务所)

合伙人 partners: Markus Kaasik · Ilmar Valdur · Andres Ojari
Raul Kalvo · Juhan Rohtla · Tanel Tatsi · Pirko Vomma

塔林是波罗地海地区经济发展最为快速的国家之一——爱沙尼亚的首都。在过去的十年里，塔林已发展成为一个现代化程度很高的开放城市。新的行政大楼将占地35000平方米，毗邻举办音乐会和体育赛事的Linnahall。**该竞赛引起了多方关注，吸引了81位建筑师及他们的设计团队参与竞赛。**国际评审小组从中选出9个最佳方案的设计团队作为决赛选手参与第二轮的竞赛。

Tallinn is the capital of one of the most rapidly developing economies in the Baltic Sea Region. During the last 10 years it has developed into a very modern and open city. The new administrative building will be located on a 35,000 m2 plot close to the concert/sports venue called Linnahall. **The contest was met with a great interest, 81 architects presented an entry.** Of those, the international jury chose the best 9 to shortlist as finalists into the second phase of the competition.

市府新行政大楼 · 塔林

New administrative building of the city Government · Estonia

一等奖 · First Prize

the public village (竞标代码)

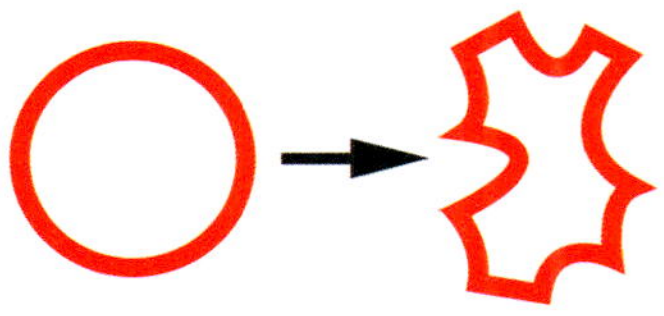

BIG · Bjarke Ingels Group (建筑师事务所)

- 教育局 EDUCATION DEPARTMENT
- 环境局 ENVIRONMENT DEPARTMENT
- 市政局 CITY ENTERPRISE DEPARTMENT
- 市政警察局 MUNICIPAL POLICE DEPARTMENT
- 住房经济局 HOUSING ECONOMY DEPARTMENT
- 城市规划局 CITY PLANNING DEPARTMENT
- 文化遗产局 CULTURAL HERITAGE DEPARTMENT
- 土地资源局 LAND ISSUES DEPARTMENT
- 市政工程局 MUNICIPAL ENGINEERING DEPARTMENT
- 社会福利及医保局 SOCIAL WELFARE AND HELTHCARE DEPARTMENT
- 交通局 TRANSPORT DEPARTMENT
- 市政厅 COUNCIL
- 主要流通 MAIN CIRCULATION

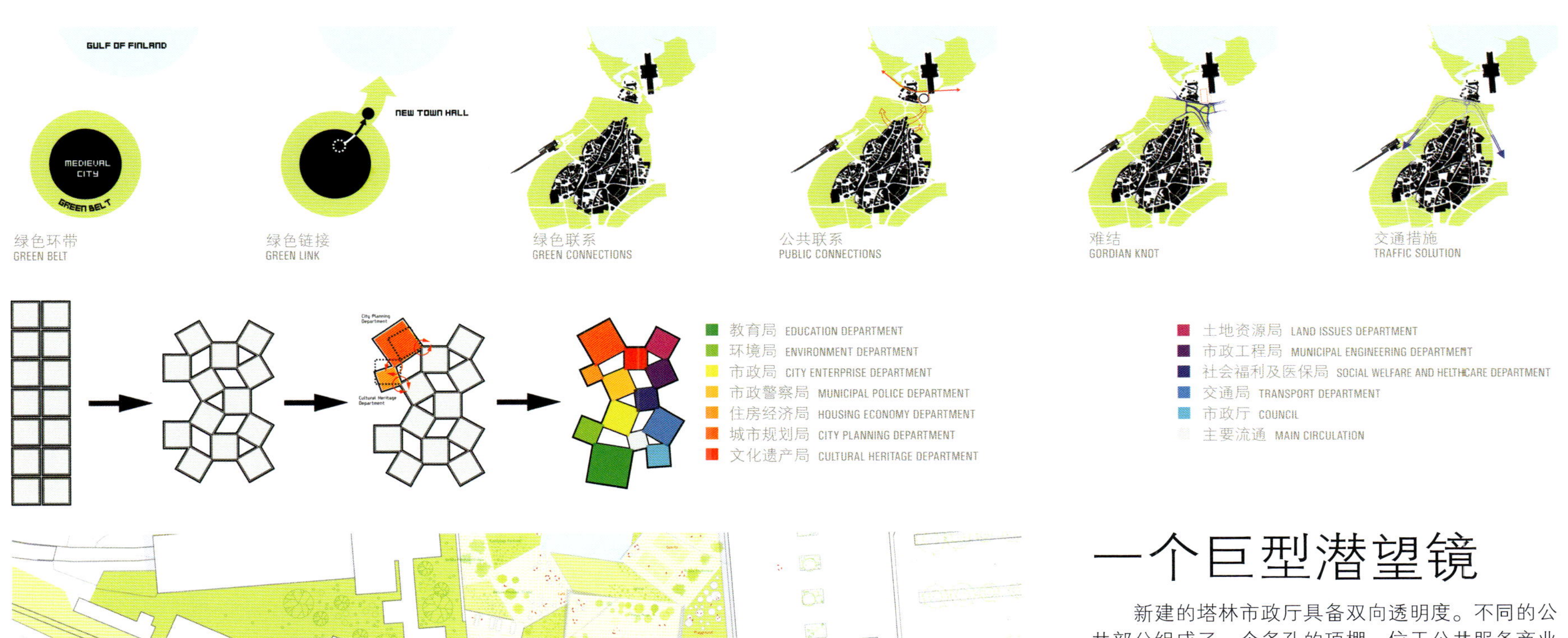

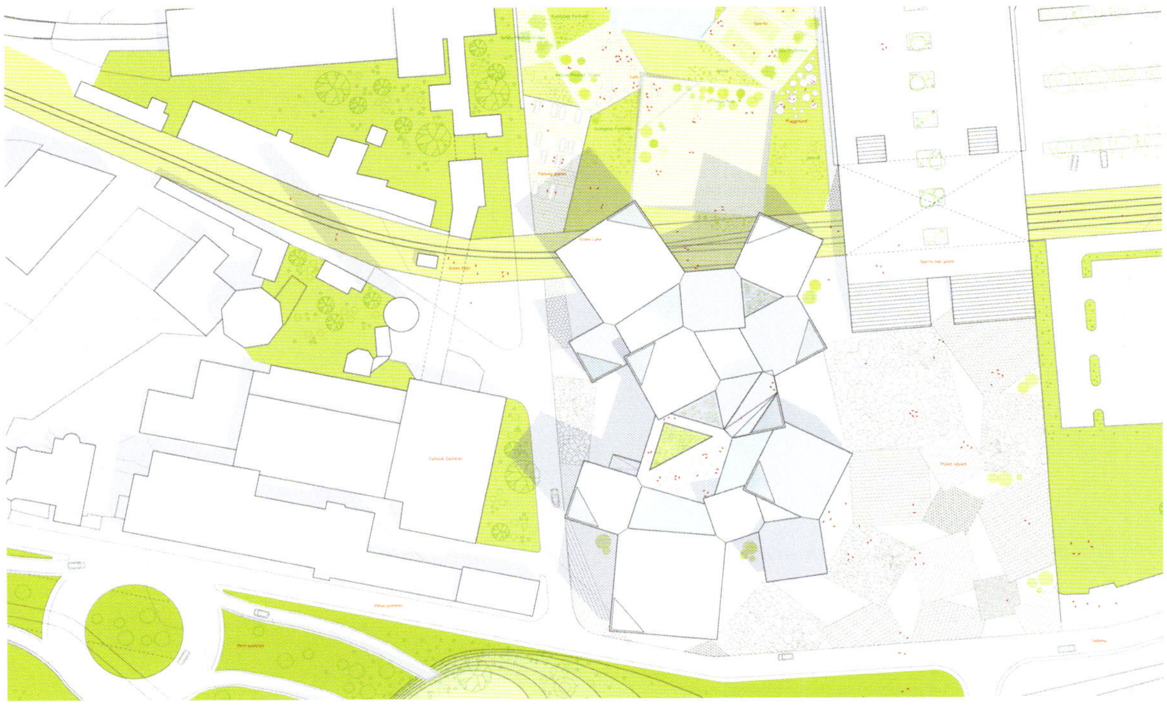

一个巨型潜望镜

新建的塔林市政厅具备双向透明度。不同的公共部分组成了一个多孔的顶棚，位于公共服务商业中心地上方，使照明可采自然光，且能饱览建筑内景。市议会位于市政厅大厦中，可鸟瞰公园和广场美景。大厦的屋顶倾斜而建，呈尖塔状。大厦倾斜的天花板由反射材料润饰，看上去仿若一个巨型潜望镜。

A HUGE PERISCOPE

The new town hall of Tallinn provides a two-way transparency in a very literal way. The various public departments form a porous canopy above the public service market place allowing both daylight and view to permeate the structure. The City Council is located in the town hall tower visible from the park and the plaza. The roof of the tower is tilted forming a slender spire. The sloping ceiling of the tower is finished in a reflective material as a huge periscope.

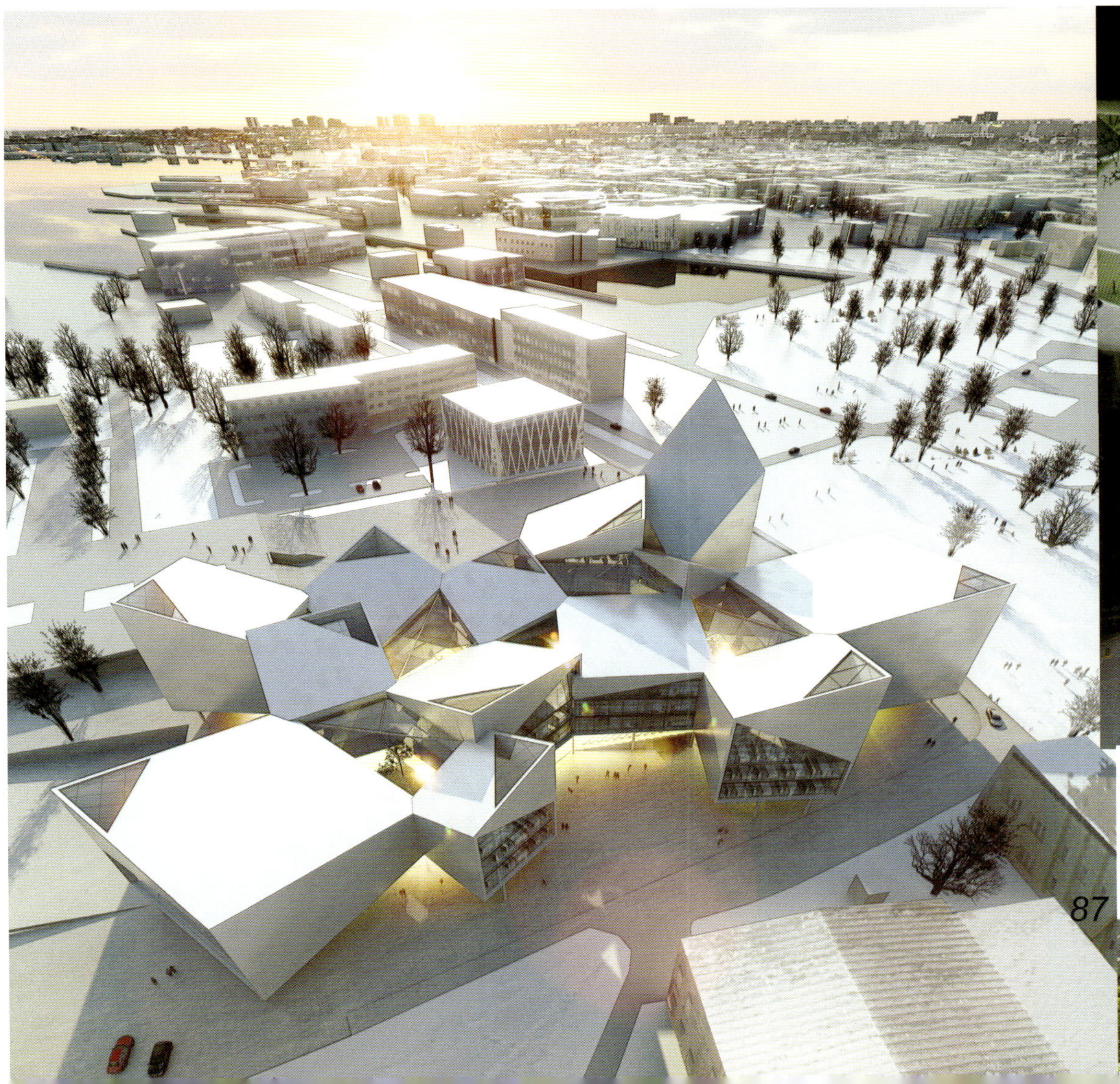

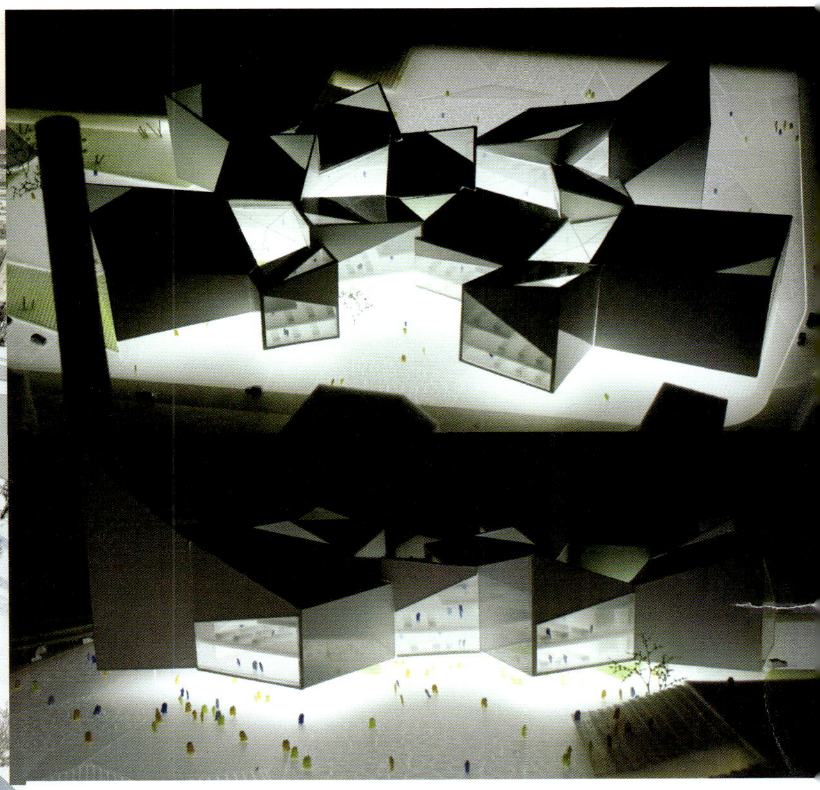

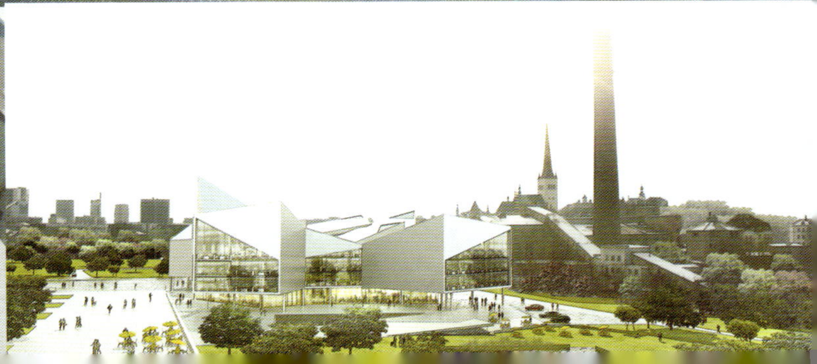

二层平面图 FIRST FLOOR PLAN

四层平面图 THIRD FLOOR PLAN

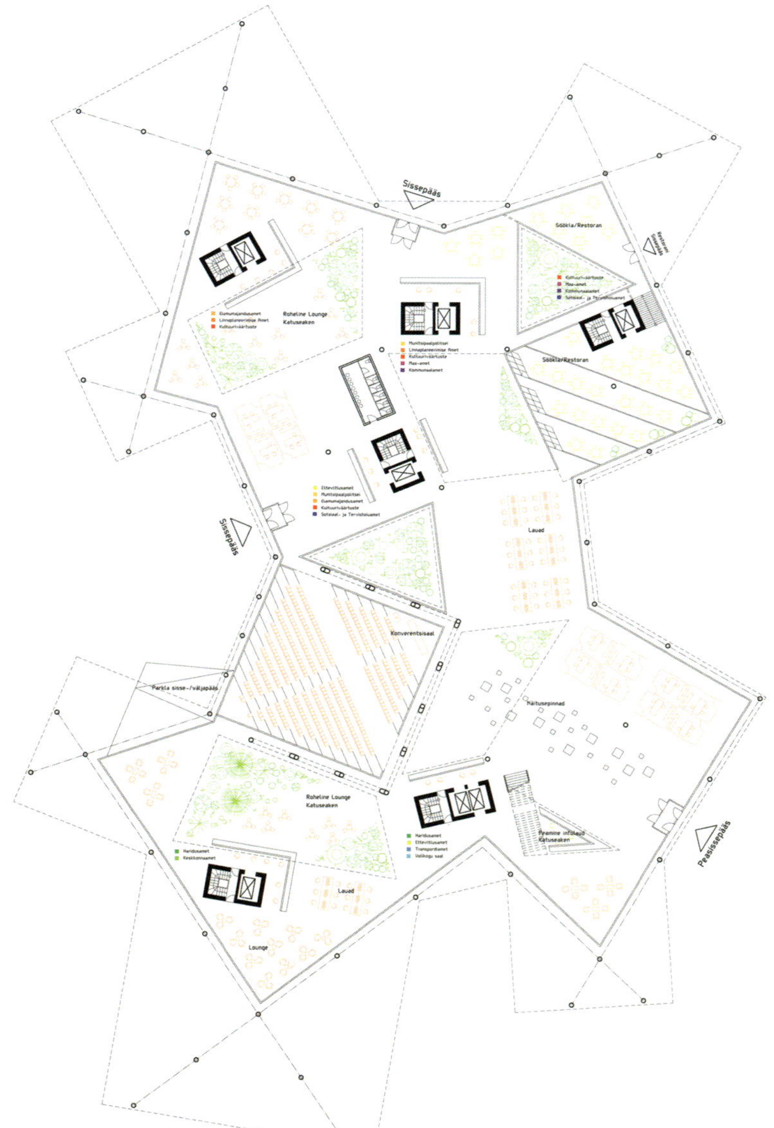

底层平面图 GROUND FLOOR PLAN

三层平面图 SECOND FLOOR PLAN

剖面图 AA SECTION AA

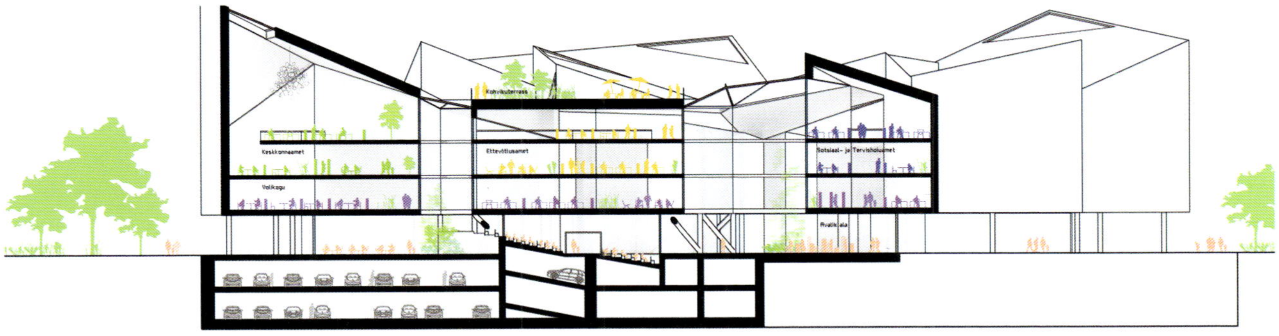

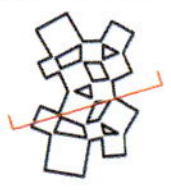

剖面图 BB SECTION BB

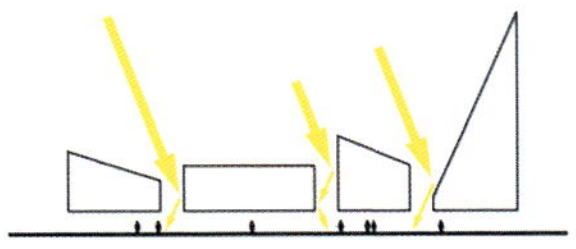

天井 COURTYARDS

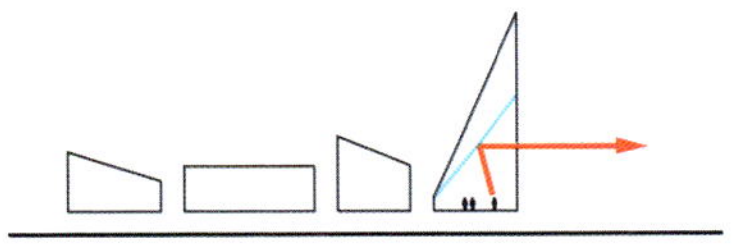

总体视线 OVERVIEW

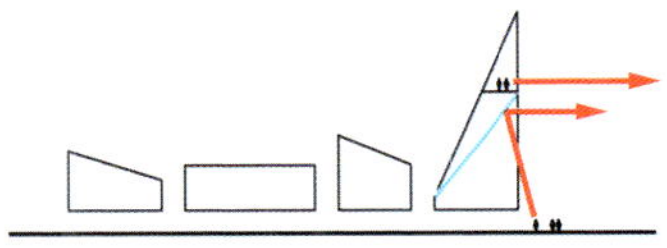

民主视线 DEMOCRATIC VIEW

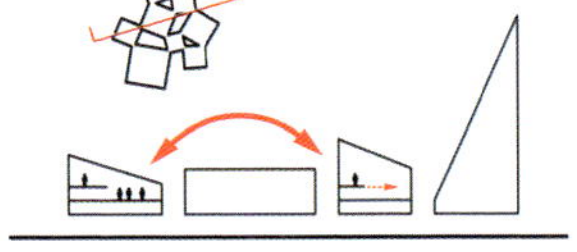

特殊灵活性 SPATIAL FLEXIBILITY

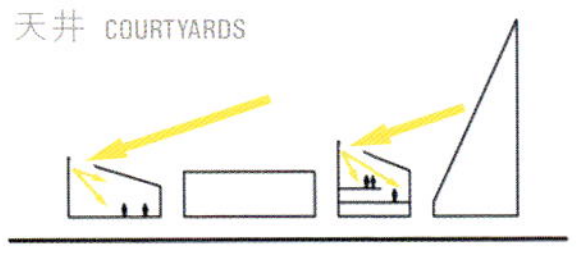

露天日照 SKYLIGHTS

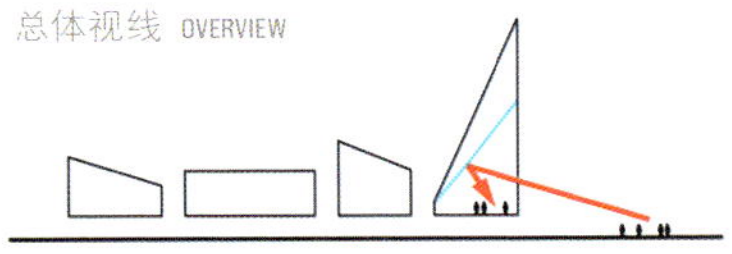

内部视线 INSIGHT

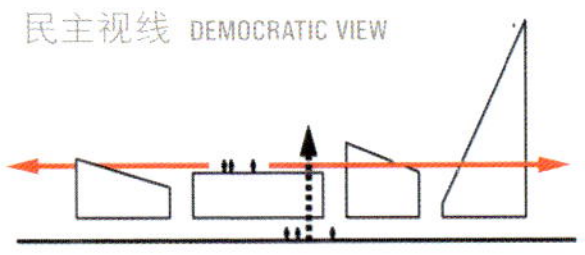

公共露台 PUBLIC ROOF TERRACE

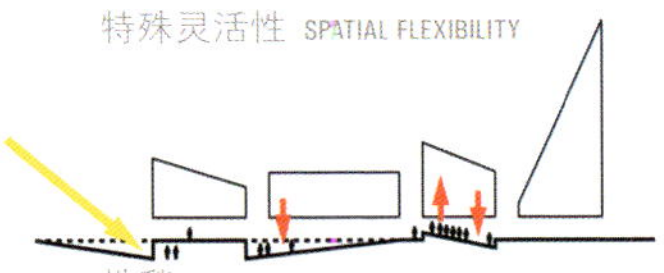

地毯 THE CARPET

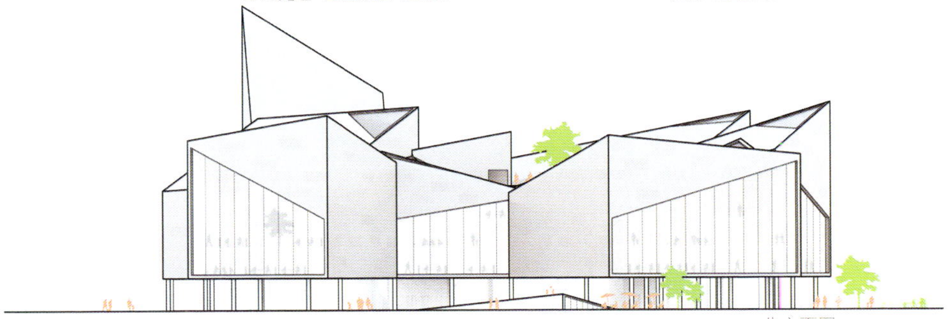

北立面图 NORTH ELEVATION

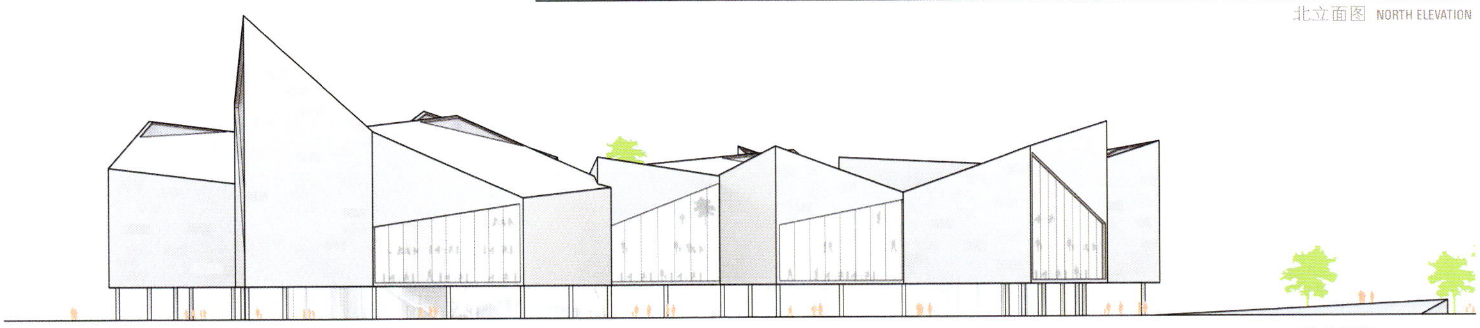

东立面图 EAST ELEVATION

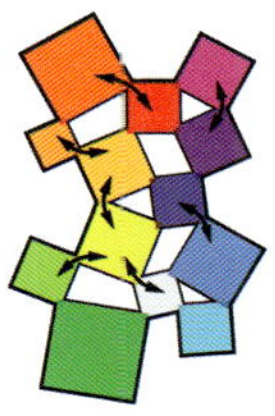

连接 CONNECTIONS

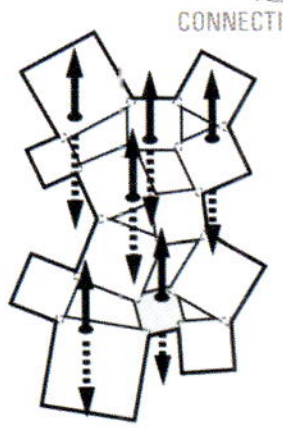

电梯核心 ELEVATOR CORES

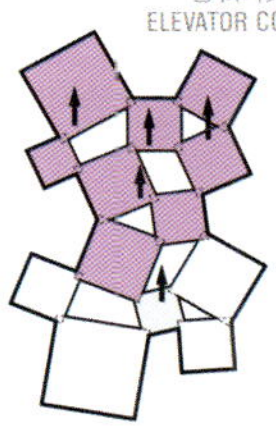

城市办公室 CITY OFFICE

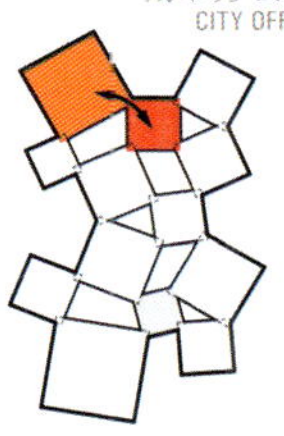

部门间连接 DEPARTMENT CONNECTION

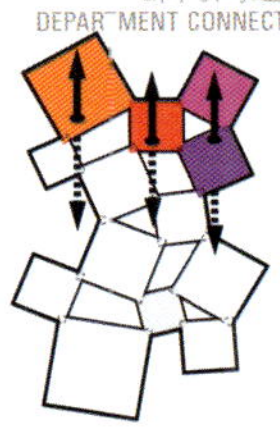

档案电梯 ARCHIVE ELEVATORS

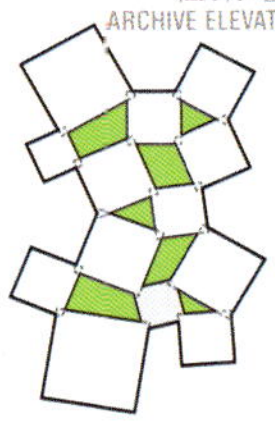

天井 COURTYARDS

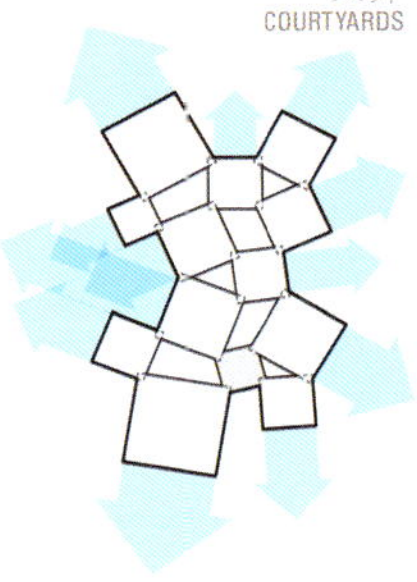

全景视线 PANORAMIC VIEWS

市府新行政大楼 · 塔林

New administrative building of the city Government · Estonia

二等奖 · Second Prize

rahva hääl (竞标代码)

Alver Arhitektid (建筑师事务所)

开放广场 > 带顶广场

这幢新建筑的功能是要将老城区和待开发的海滨地带结合起来。根据这一方案，市政厅前空出了一些连续的公共空间。这幢新建筑由两部分组成：大厅（议会和餐厅）和各部门的办公楼。

OPEN SQUARE > COVERED SQUARE

The new building has to bind together the Old Town and seaside potential development area. The solution creates a series of continuous public spaces in front of the Town Hall. The building is composed of two parts; the foyer (the Council and the restaurant) and the set of offices housing the departments.

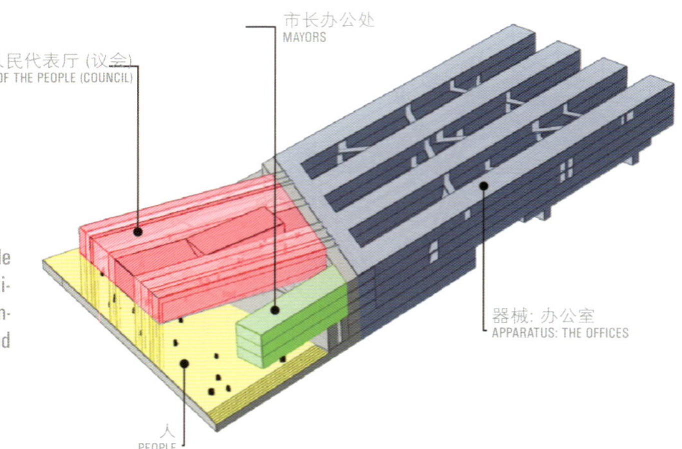

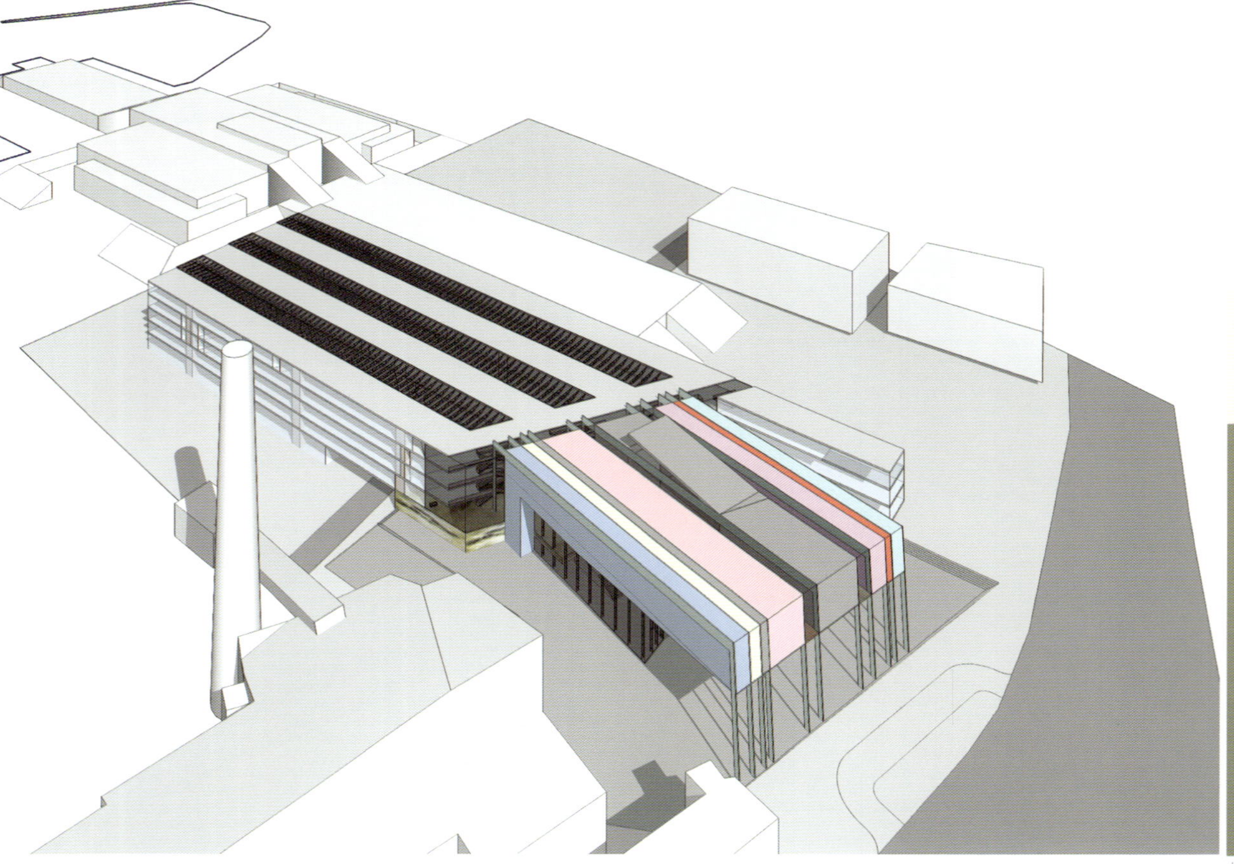

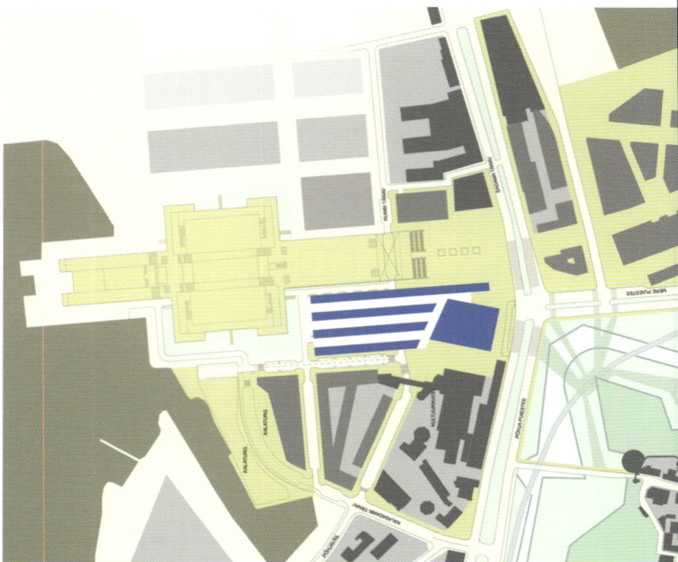

地块位置 SITE PLAN

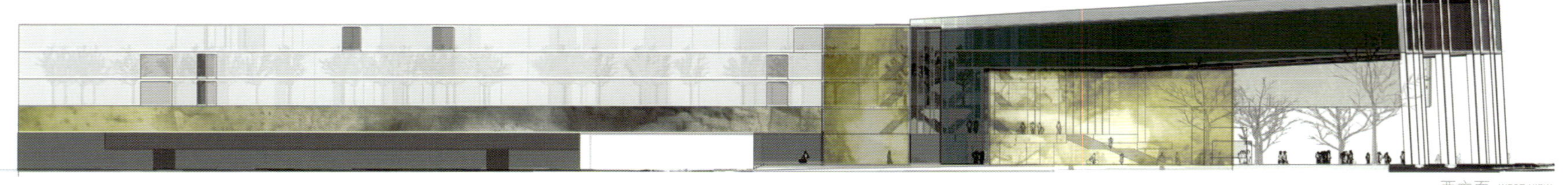

西立面 WEST VIEW

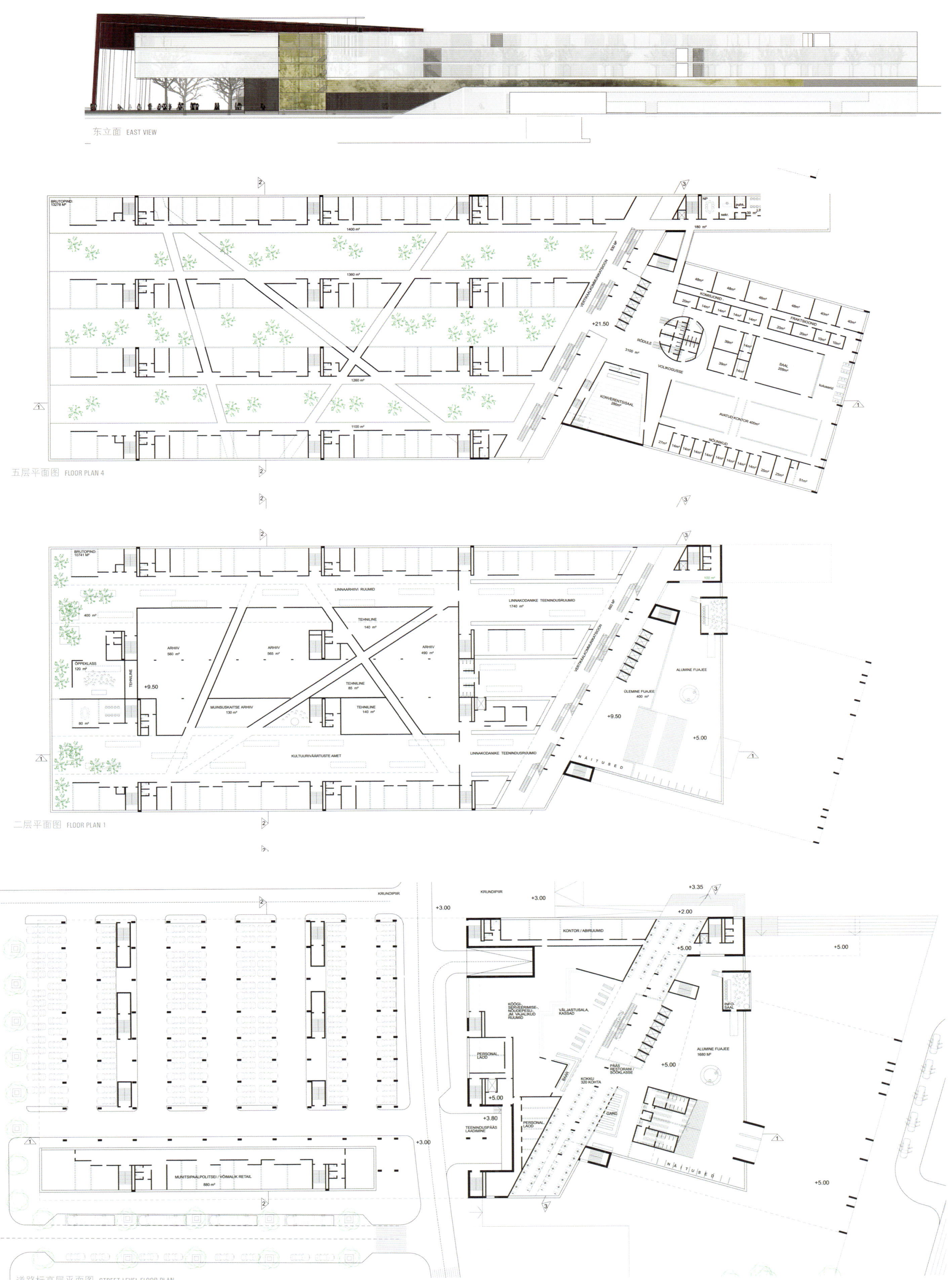
东立面 EAST VIEW

五层平面图 FLOOR PLAN 4

二层平面图 FLOOR PLAN 1

道路标高层平面图 STREET LEVEL FLOOR PLAN

市府新行政大楼 · 塔林

New administrative building of the city Government · Estonia

三等奖 · Third Prize

koda (竞标代码)

3+1 architects (建筑师事务所)

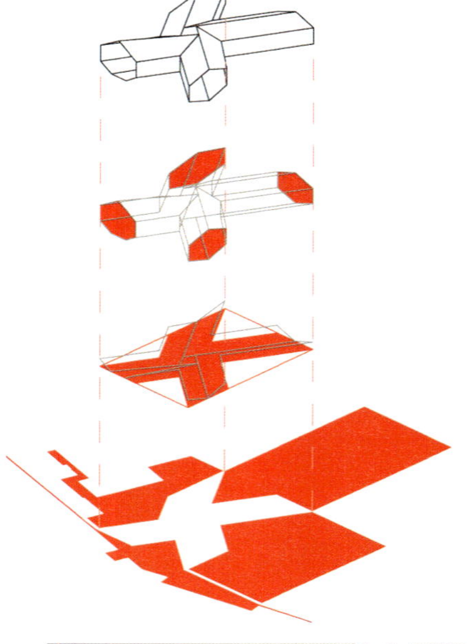

原型

在大海和旧城区之间重新打造“铺设的沙漠”，将会巧妙地在空间上分出主干道和城市街道。新建的市政厅与各个重要的城市综合体在文化氛围上相互契合，如举办音乐会和体育赛事的巨型建筑Linnahall以及一个设在名为Kultuurikatel陈旧发电站里的活动中心。这样做的目的在于将这些分开的机构整合到一个城市广场中。

ARCHETYPE

Reorganizing the "paved desert" between the sea and the Old Town will create a tactical solution with the objective to spatially define the identities of street-types: artery and city-street. The new City Hall links culturally important complexes: the concert/sports megastructure venue Linnahall and an event center in an old powerplant named Kultuurikatel. The aim is to connect these separate institutions into an urban plaza.

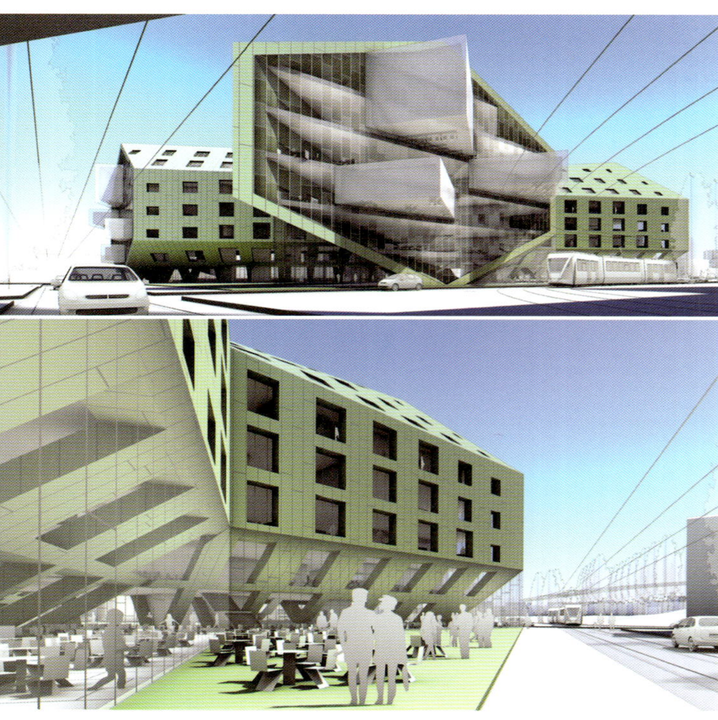

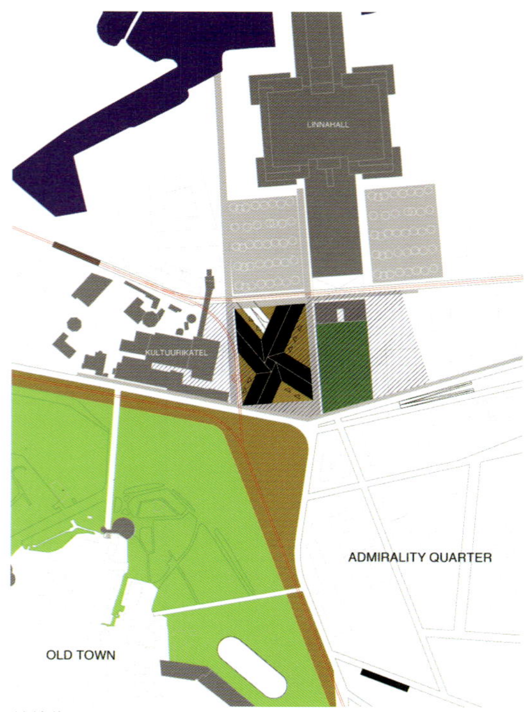

地块位置 SITE PLAN

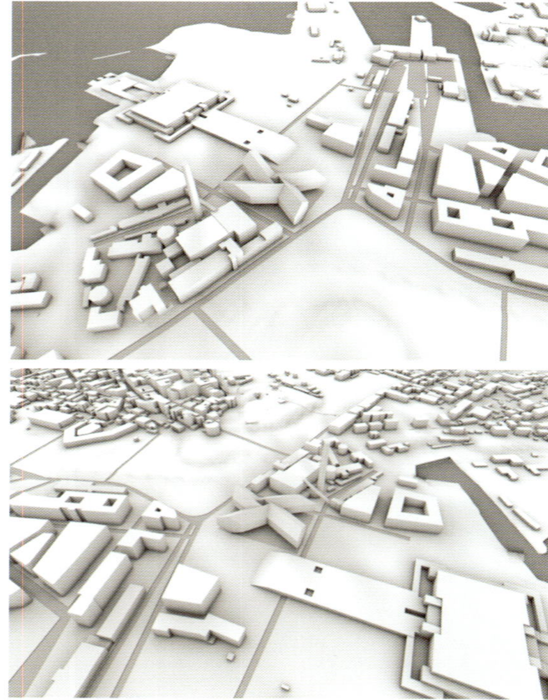

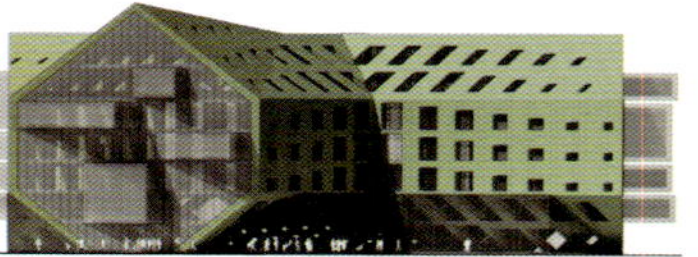

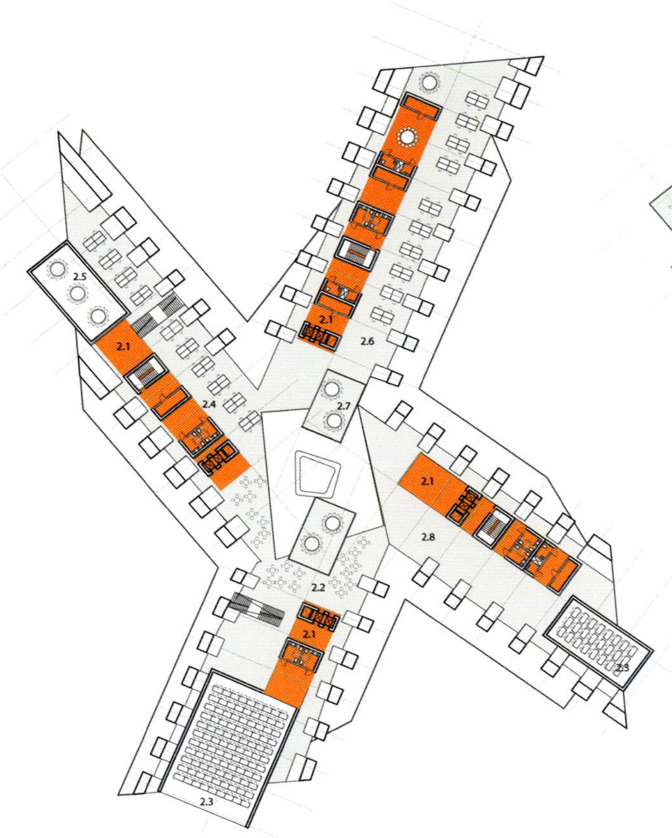

三层平面图 FLOOR PLAN 2

四层平面图 FLOOR PLAN 3

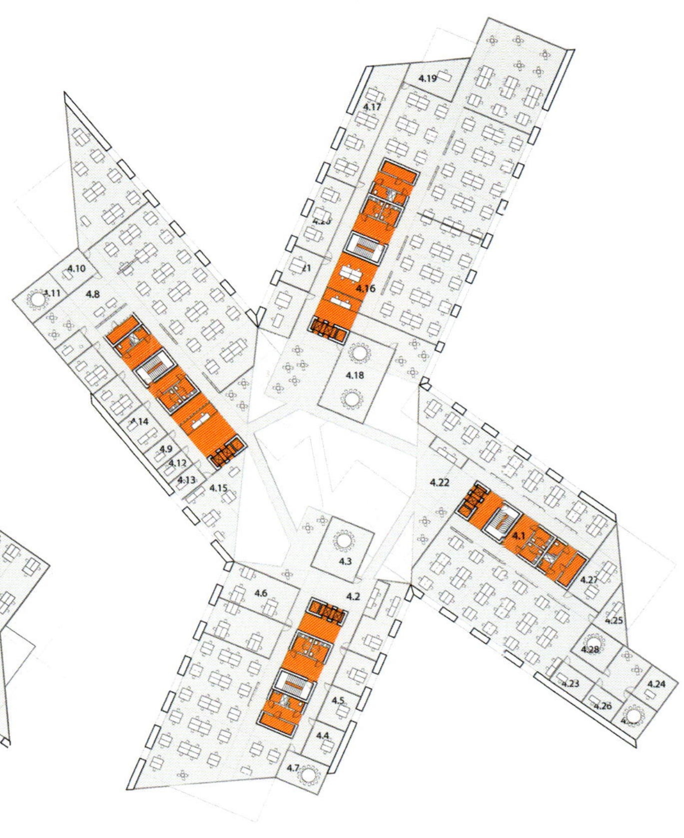

五层平面图 FLOOR PLAN 4

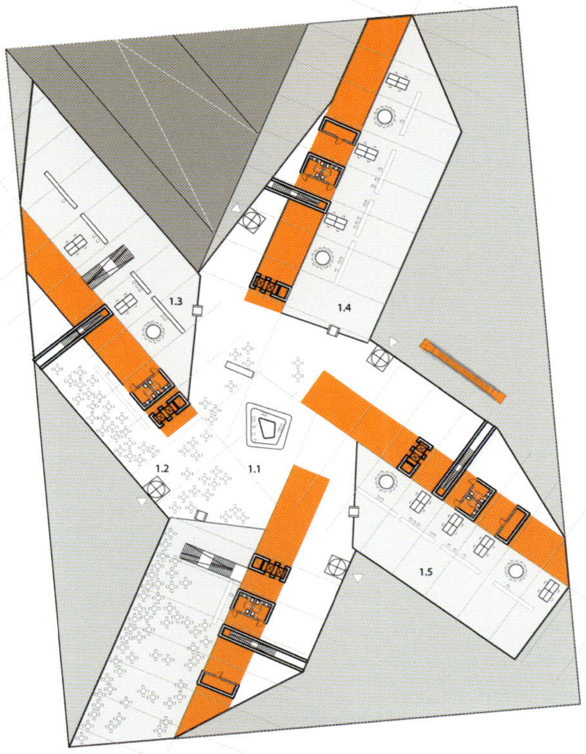

底层平面图 GROUND FLOOR PLAN

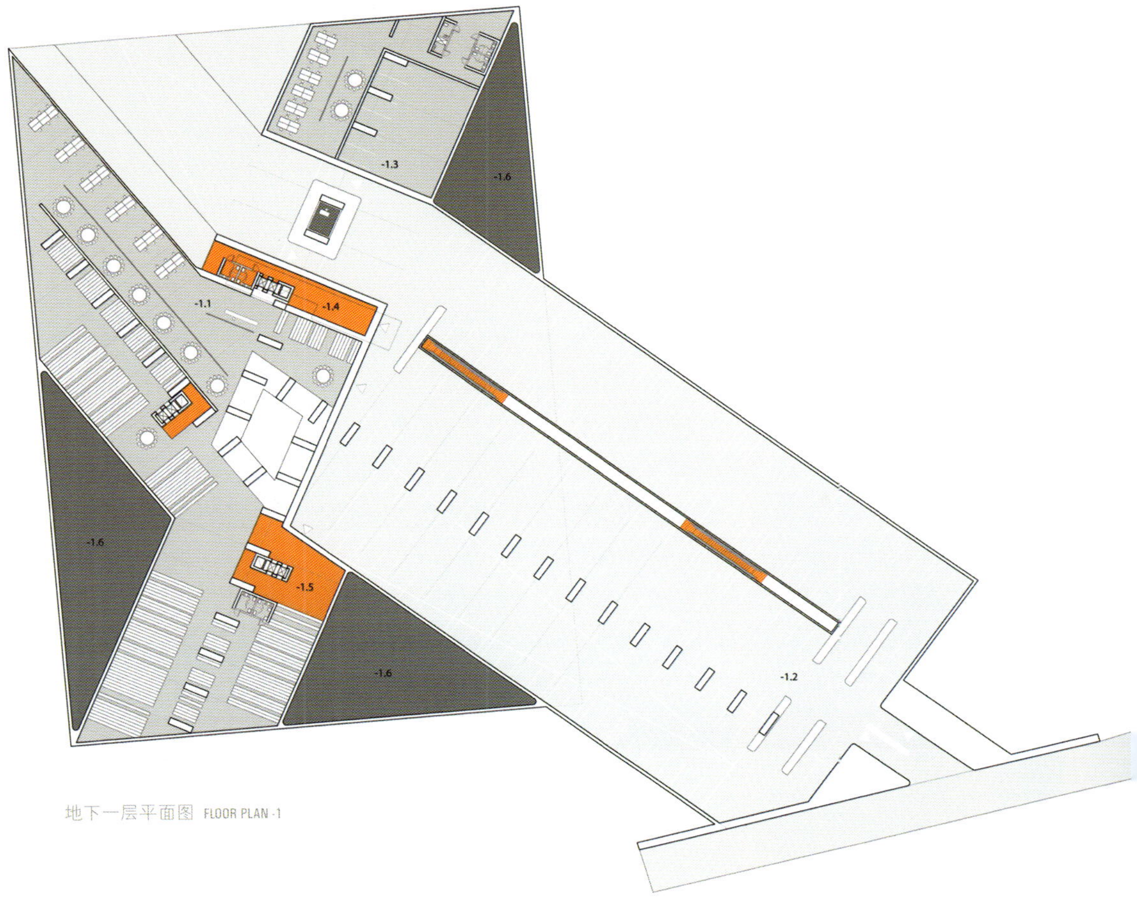

地下一层平面图 FLOOR PLAN -1

经济与企业大学新校区（WU）· 维也纳
New Campus of the Vienna University of Economics and Business (WU) · Austria

竞标 · competition
维也纳经济与企业大学新校区（WU）
New Campus of the Vienna University of Economics and Business (WU)

竞标类型 · competition type
两阶段竞标，一阶段六名参标者，二阶段三名参标者
Two-phase competition with six competitors in the first phase and three in the second phase

项目地点 · site area
维也纳 · 奥地利 Viena · Austria

主办方 · promoter
维也纳经济与企业大学及Bundesimmobiliengesellschaft公司，联合奥地利联邦科学研究部部长Johannes Hahn及市议员Rudolf Schicker（城市发展与交通部）
The WU and Bundesimmobiliengesellschaft (BIG), together with Austria's Federal Minister of Science and Research Johannes Hahn and City Councillor Rudolf Schicker (Department for Urban Development and Traffic)

日程安排 · schedule
招标 · Announcement 06.2008
评审结果 · Jury's results 12.2008

评审团 · jury

Wolf D. Prix · 主席 · chairman
Dietmar Eberle · 副主席 · deputy chairman
Brian Cody
Peter Ehrenberger
Laura P. Spinadel
Sepp Frank
Klaus Vatter
Christoph Badelt
Sepp Hannreich
Michael Holoubek
Gerhard Kubik
Günther Sokol · 秘书 · secretary

获奖者 · awards

总建筑项目 General Construction Project

中标 · winner
BUSARCHITEKTUR & PARTNER（建筑师事务所）

二等奖 · second prize
ARGE FLATZ_ARCHITECTS + ZEYTINOGLU ARCHITECTS（建筑师事务所）

三等奖 · third prize
PROF. HANS HOLLEIN（建筑师事务所）

部门大楼(02) Department Building (02)

中标 · winner
ATELIER HITOSHI ABE（建筑师事务所）

项目经理 · project manager · Pierre De Angelis
项目团队 · project team · Regina Blauensteiner, Carmen Cham, Tim Do, Jon Ford, Midori Mizuhara, Mina Nishio, Atsushi Sugiuchi, Joe Willendra, Kentaro Yamada

二等奖 · second prize
BEVK PEROVIC ARHITEKTI（建筑师事务所）
团队 · team · Matija Bevk, Vasa J. Perović, Anja Vidic, Tadej Kališnik, Andrej Ukmar, Ida Sedušak, Gerrit Neumann, Blaz Goričan, Blaz Hartman, Vanja Otasevic, Marija Krsmanovic

三等奖 · third prize
JOSEP LLINAS I CARMONA（建筑师事务所）

图书馆及学习中心(LLC) Library & Learning Center (LLC)

中标 · winner
ZAHA HADID ARCHITECTS（建筑师事务所）

设计 · design · Zaha Hadid and Patrik Schumacher
项目 · project · Architect Cornelius Schlotthauer
项目团队 · project team · Marc-Philipp Nieberg, Kristoph Nowak, Enrico Kleinke, Stefan Rinnebach, Niels Kespohl, Jan Hübener, Romy Heiland, Richard Baumgartner
结构工程 · structural engineers · Arup Berlin
机械工程 · M&E engineers · Arup Berlin
幕墙工程 · façade engineers · Arup Berlin
成本预算 · cost consultant · ATP Wien
防火工程 · fire protection · HHP West, Bielefeld
渲染 · render studio · Vectorvision, Leipzig

二等奖 · second prize
THOM MAYNE - MORPHOSIS（建筑师事务所）

三等奖 · third prize
PROF. HANS HOLLEIN（建筑师事务所）

部门大楼(W1) Department Building (W1)

中标 · winner
ESTUDIO CARME PINÓS（建筑师事务所）
团队 · team · Holger Hennefarth, Roberto Carlos García, Juan Antonio Andreu, Alexandra Clausen

结构 · structure · Robert Brufau · BOMA
模型 · model · Estudio Carme Pinós

二等奖 · second prize
NO.MAD ARQUITECTOS（建筑师事务所）

三等奖 · third prize
HOLZBAUER + PARTNER（建筑师事务所）

部门大楼(W2) Department Building (W2)

中标 · winner
CRABSTUDIO（建筑师事务所）
Peter Cook · Gavin Robotham

二等奖 · second prize
GUILLERMO VÁZQUEZ CONSUEGRA（建筑师事务所）

团队 · team · Juan Jose Baena (coordinador), Cristoph Haralter, Kristian de Solaun, Marko Jovanovic, Filippo Pambianco, Samuele Evolvi, Alessandro Tessari, Joanna Jedrus, Pedro Hébil

三等奖 · third prize
ERIC OWEN MOSS ARCHITECTS（建筑师事务所）

团队 · team · Eric Owen Moss, Eric McNevin, Jose Herrasti, Fernanda Bento
BURO HAPPOLD Consulting Engineers, Los Angeles

维也纳经济管理大学现有在校学生23000人，校园里十分拥挤，不利于师生们的学习、工作和研究。这一情况将于2013年转变。**新校园位于普拉特和Messe博览会展览中心之间的幽静之处**，是城市结构的一部分，足够的开放空间将会彰显都市校园的理念。

BUSarchitektur制订的总体规划包括基础设施和自由空间的设计，并且将该工程分为几个部分。竞争主题是根据总体规划来设计五个建筑综合体，连同BUSarchitektur设计的演讲教室中心和院系楼(01)。这个占地90000平方米的场地上将建起风格各异的建筑，建筑面积将达到102000平方米。

With its present roster of some 23,000 students, the Vienna University of Economics and Business is quite bursting, which does not make it easy to study, to work and to research. That will change in 2013. **The new WU Campus site lies hidden away between the Prater and the Messe Fair exhibition Centre**, embedded in a city structure with sufficient open space to develop an urban campus concept.

The Master Plan by BUSarchitektur laid down the infrastructure and free space design and divides the project into several plots. The subject of the competition was the design of five building complexes on the basis of the Master Plan. Together with the Lecture Room Center and Department Building (01) designed by BUSarchitektur, the 90,000 m2 site will be characterised by a great diversity of architecture and provide the University with 102,000 m2 of floor area.

millefeuille*: a thousand layers (竞标代码)

Atelier Hitoshi Abe (建筑师事务所)

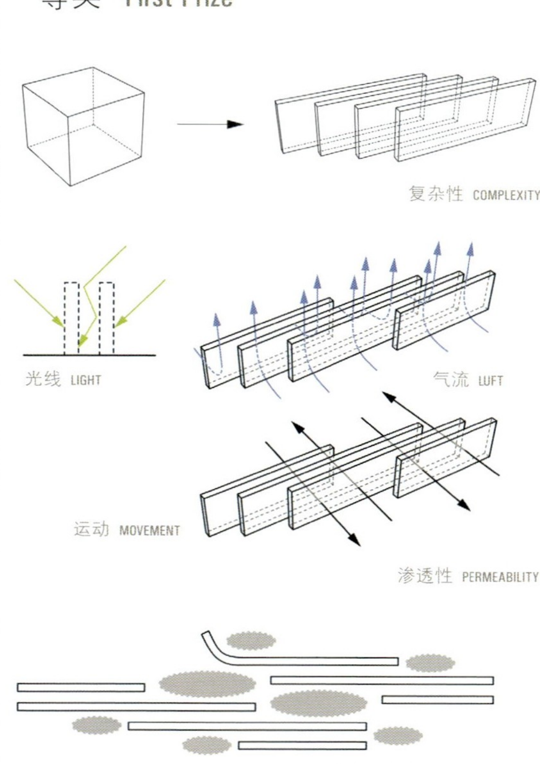

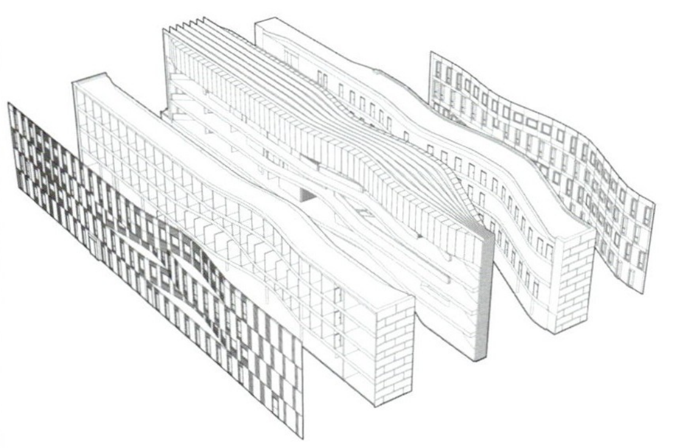

细条的分层排布使建筑物能适应不同的场地条件，因而有利于最大限度地利用自然光线和无源通风系统，并且可以满足计划性要求，为将来的调整提供极大的便利条件。

The layering of thin volumes maximizes opportunities for natural lighting and passive ventilation as it allows the building to be responsive to different site conditions, fulfil programmatic requirements and maximize flexibility for future adaptation.

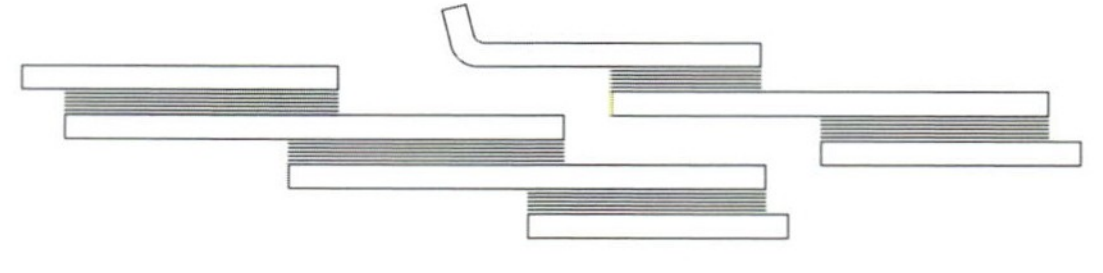

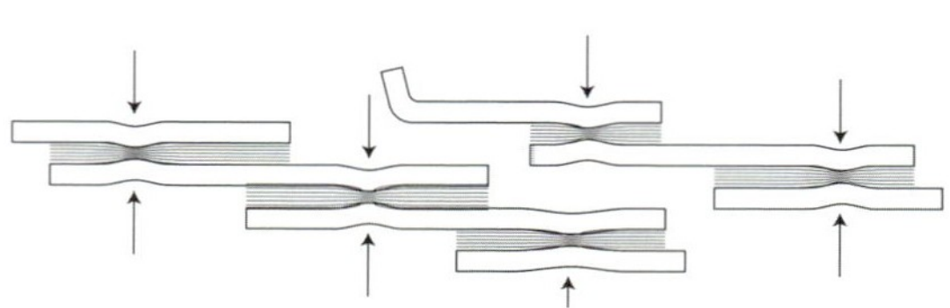

中庭中的一连串收缩结构组成了一个个形状各异的节点，人们可以从这里经过，在这里互动及聚集，造成了视觉上和空间上的差异。

A series of pinches along the length of the atriums create visual and spatial differences as it creates informal nodes where users cross paths, interact and congregate.

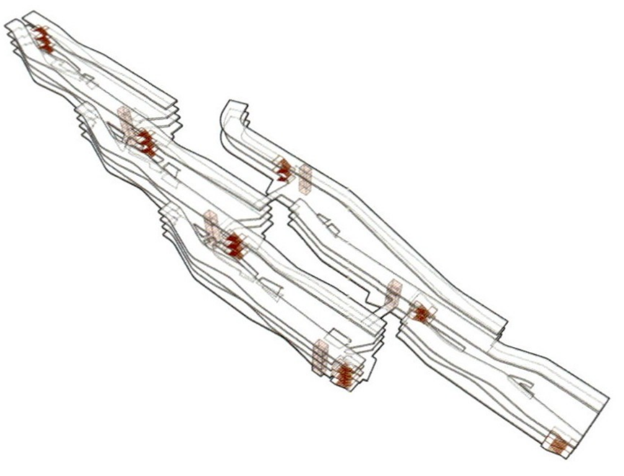

该建筑物在构造上十分统一，能在该工程使用寿命以内高度灵活地进行各种变化调节。核心机构可位于中心位置，以便各系部均能轻易到达；也可位于边缘位置，以便最大限度地满足项目的连贯性以及这些固定要素之间的灵活性。

This building is uniformly organized to allow for a high degree of flexibility in accommodating changes over the lifespan of the project. Cores are placed in either a central position; to allow for shared access between departments, or at the edges; to allow for the greatest continuity of program and maximum flexibility between these fixed elements.

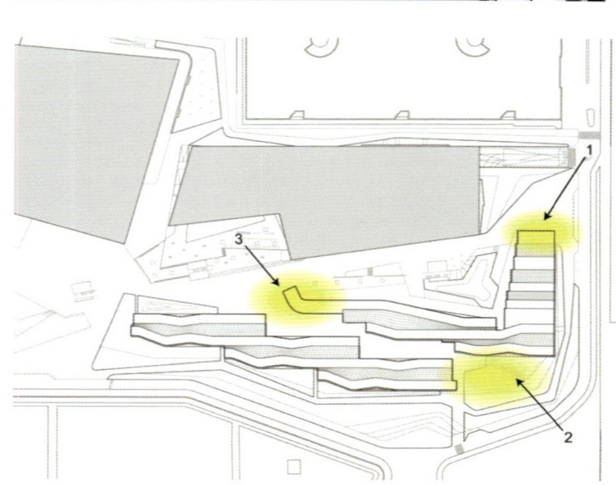

广场入口处的设计旨在欢迎广大学生、教职工以及游客的到来。

Entrances to the plazas are designed to welcome and attract students, staff, faculty and visitors.

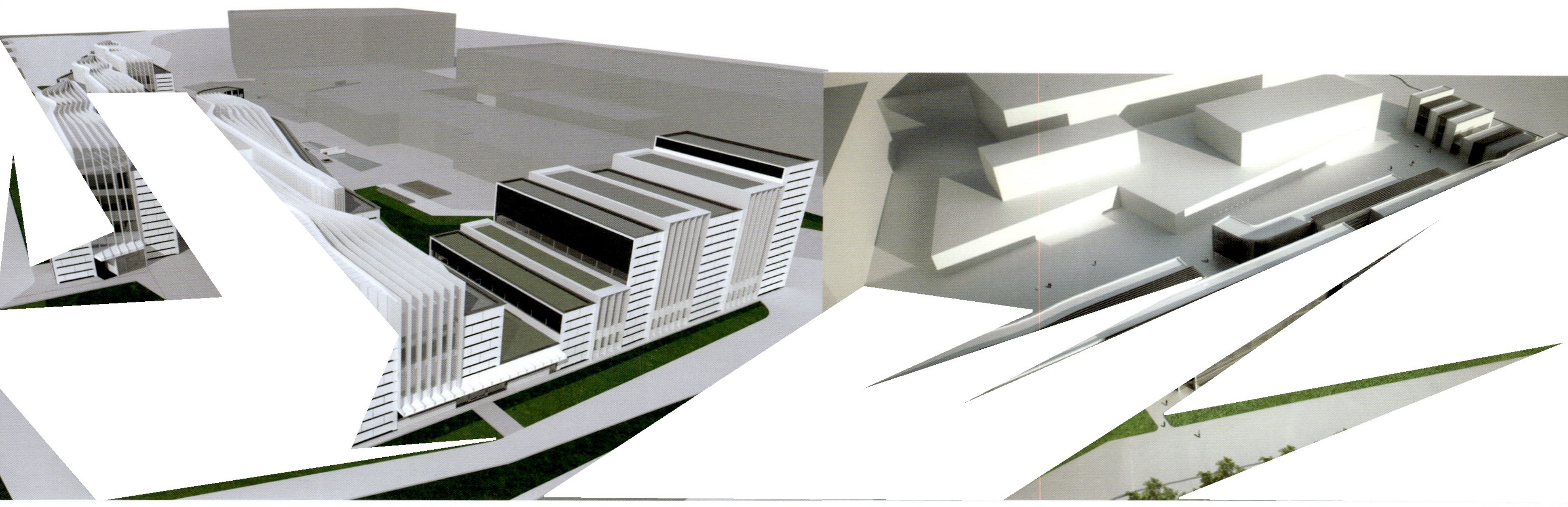

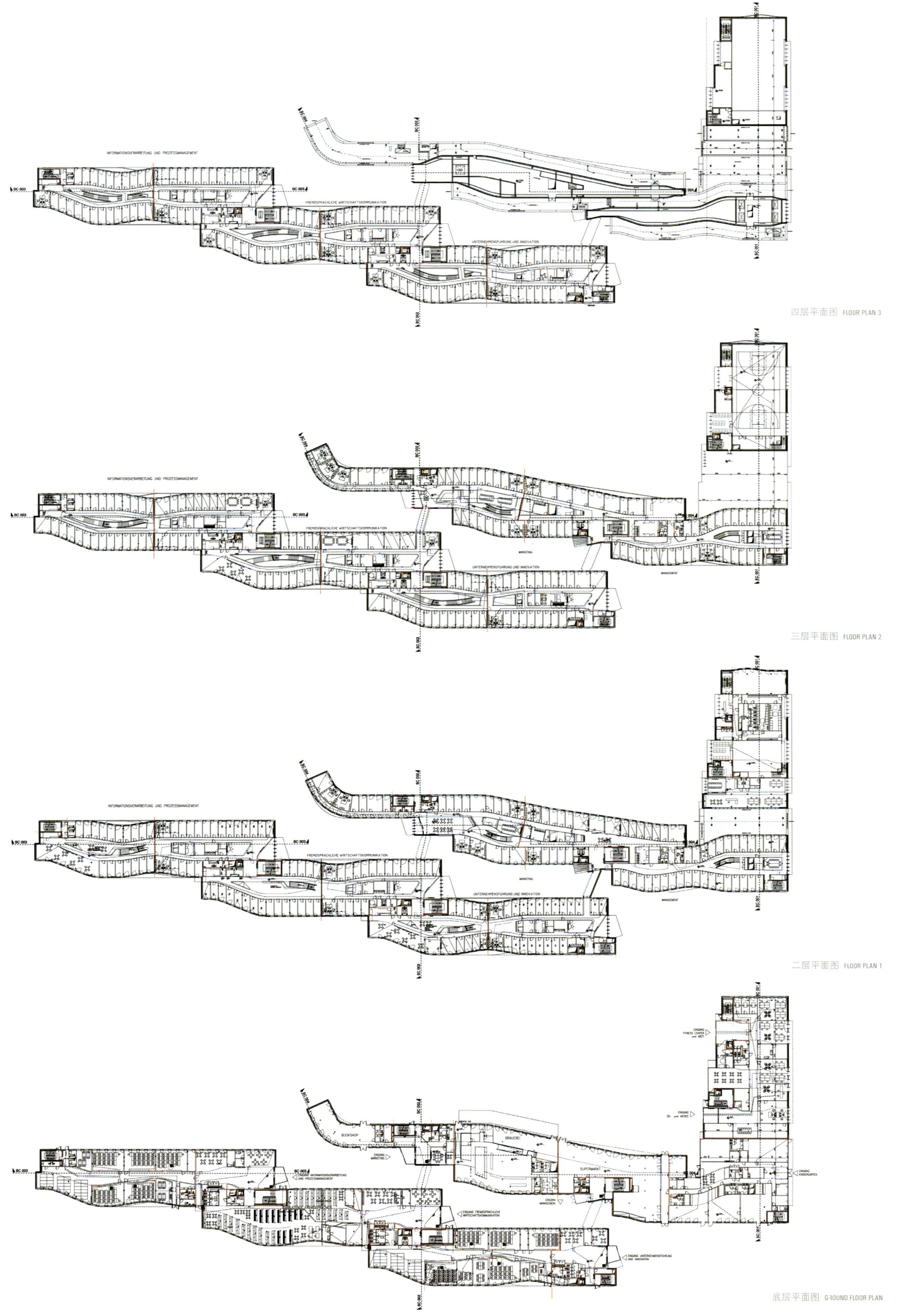

四层平面图 FLOOR PLAN 3

三层平面图 FLOOR PLAN 2

二层平面图 FLOOR PLAN 1

底层平面图 GROUND FLOOR PLAN

部门大楼 (02)

Department Building (02)

二等奖 · Second Prize

091119 (竞标代码)

Bevk Perovic arhitekti (建筑师事务所)

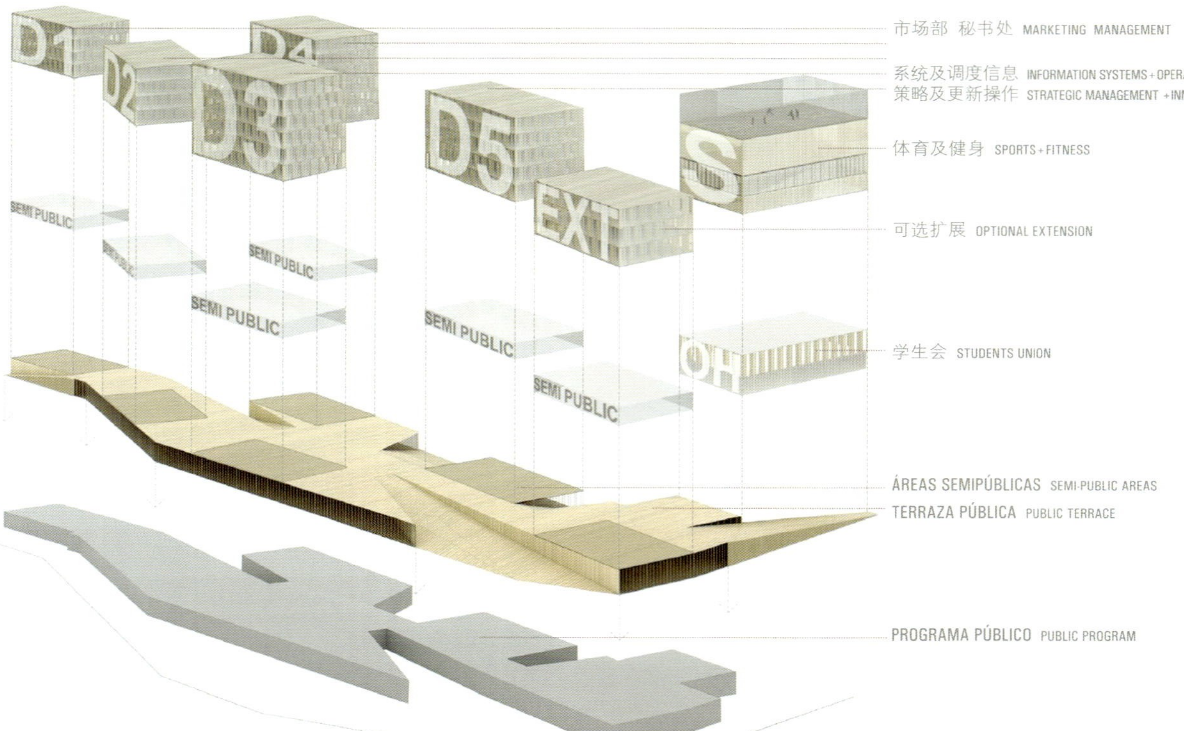

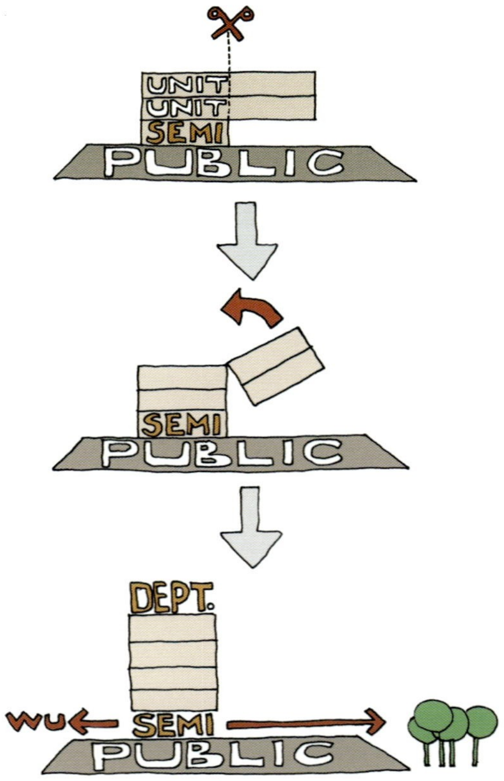

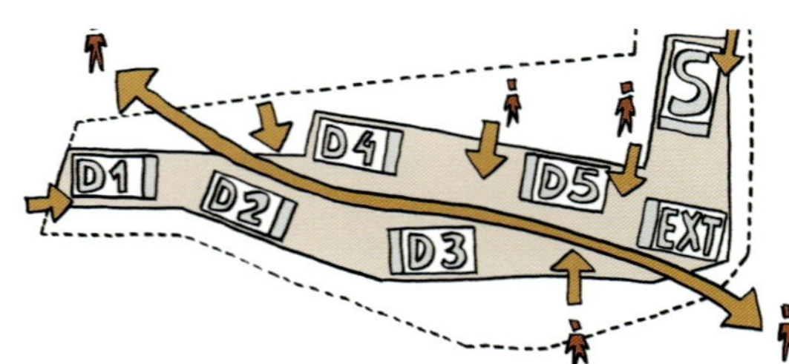

流通 CIRCULATION

PRATER + WU的视线 VIEWS PRATER + WU

建筑堆叠

新基地成了一个公共广场，一个可以俯瞰普拉特景色的“平台”。校园里又添了一道新的风景线——一个新的底层台——能使教室、学生休息室等公共设施像一个“基地”一样通往各个系部所在地。因为各系部之间无明显界限，所以每个系部的自我认同感，即将每个系部视作独立个体的可能性，变得尤为重要。

STACKS OF BUILDINGS

The new base acts as a public plaza, a 'terrace' overlooking the Prater. A new surface of the campus is established – a new ground floor – while enabling all public programs – classrooms, student lounges etc – to appear as a 'base' connected to all departments equally. Since the departments are 'mixed-up', identity becomes an important issue, i.e. the possibility to see each department as an entity.

ÖH / AIESEC
office space
tea kitchen
meeting room
office space
FOREIGN LANGUAGE & BUSINESS COMMUNICATION
meeting rooms
storage
back offices
front office
lounge
access stair
access stair
main access ramp / street
'landscape' terrace
tribune
top light
± 0,00 m (vienna 0 +3.2)
± 4,50 m
± 0,00 m
access ramp
MARKETING
green buffer
outdoor lounge
MANAGEMENT
top light
street light
INFORMATION SYSTEMS AND OPERATION
STRATEGIC MANAGEMENT AND INNOVATION
main access ramp / street
OPTIONAL EXTENSION

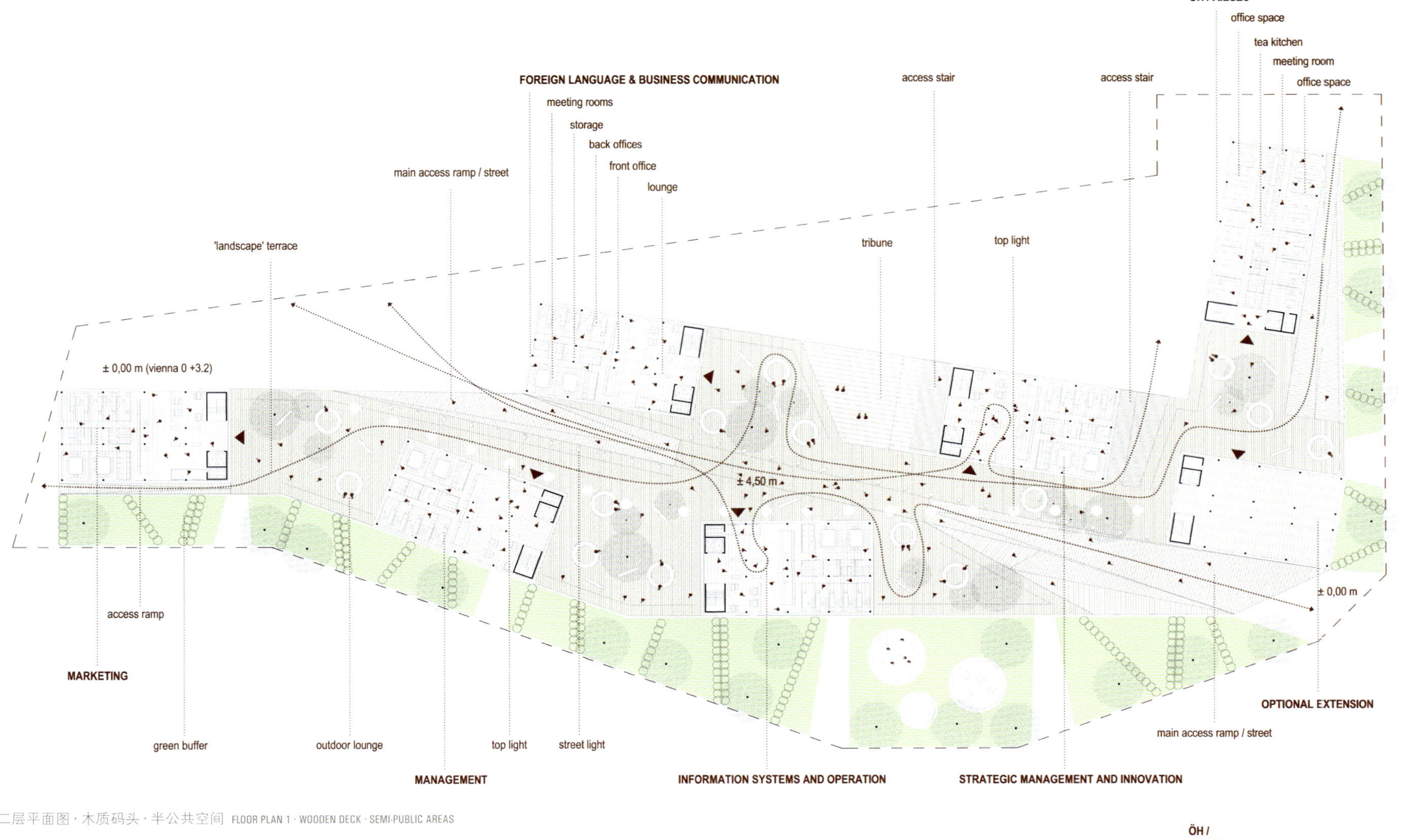

二层平面图 · 木质码头 · 半公共空间 FLOOR PLAN 1 · WOODEN DECK · SEMI-PUBLIC AREAS

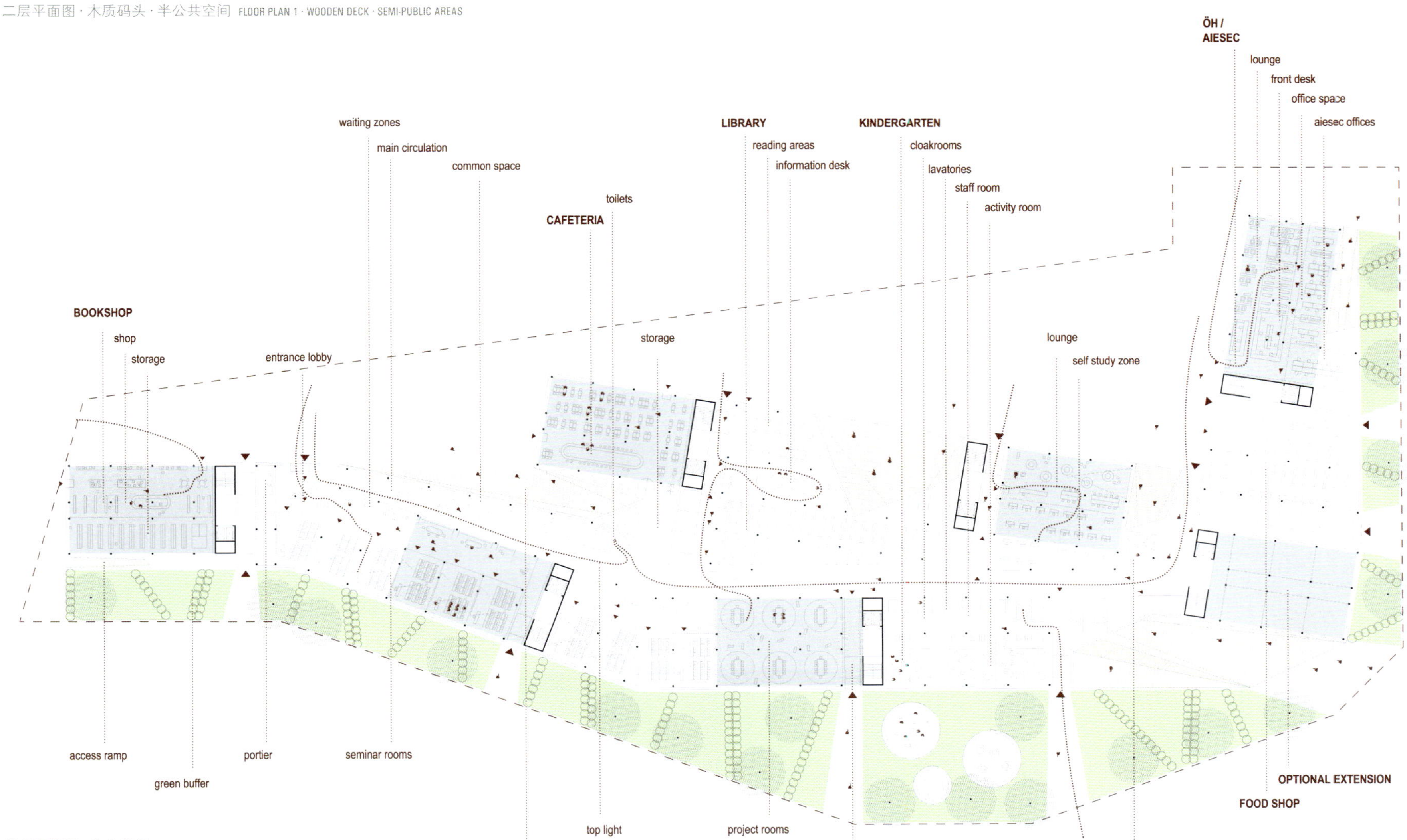

ÖH /
AIESEC
lounge
front desk
office space
aiesec offices
waiting zones
main circulation
common space
LIBRARY
reading areas
information desk
KINDERGARTEN
cloakrooms
lavatories
staff room
activity room
toilets
CAFETERIA
BOOKSHOP
shop
storage
entrance lobby
storage
lounge
self study zone
access ramp
green buffer
portier
seminar rooms
top light
project rooms
main access ramp
passage to campus
portier
OPTIONAL EXTENSION
FOOD SHOP

底层平面图 · 公共空间 GROUND FLOOR PLAN · PUBLIC AREAS

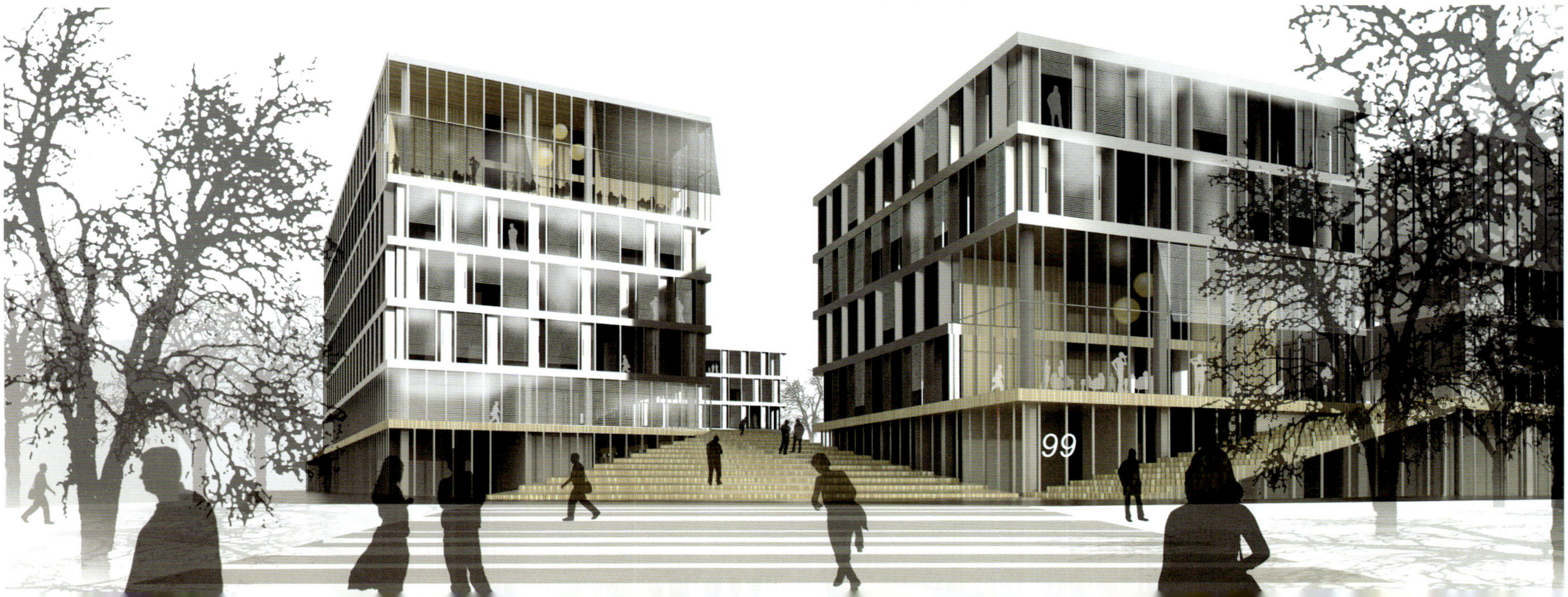

部门大楼 (02)

Department Building (02)

三等奖 · Third Prize

091119 (竞标代码)

Josep Llinas i Carmona (建筑师事务所)

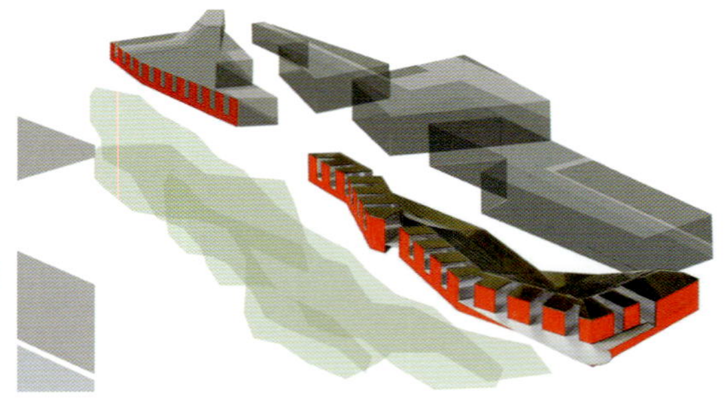

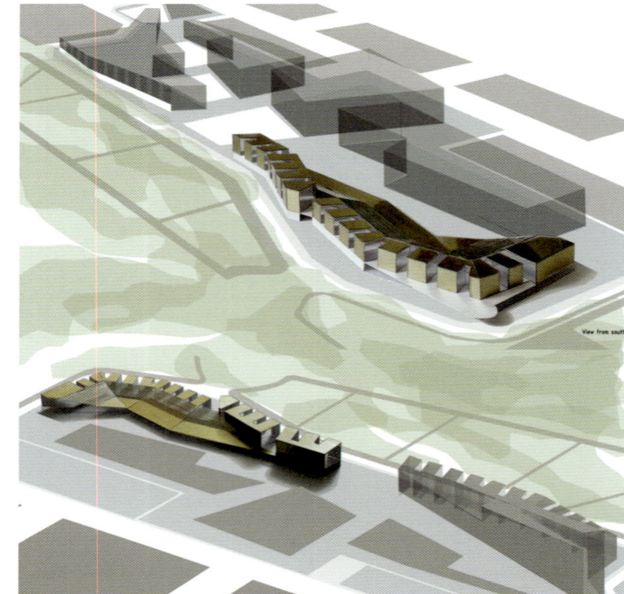

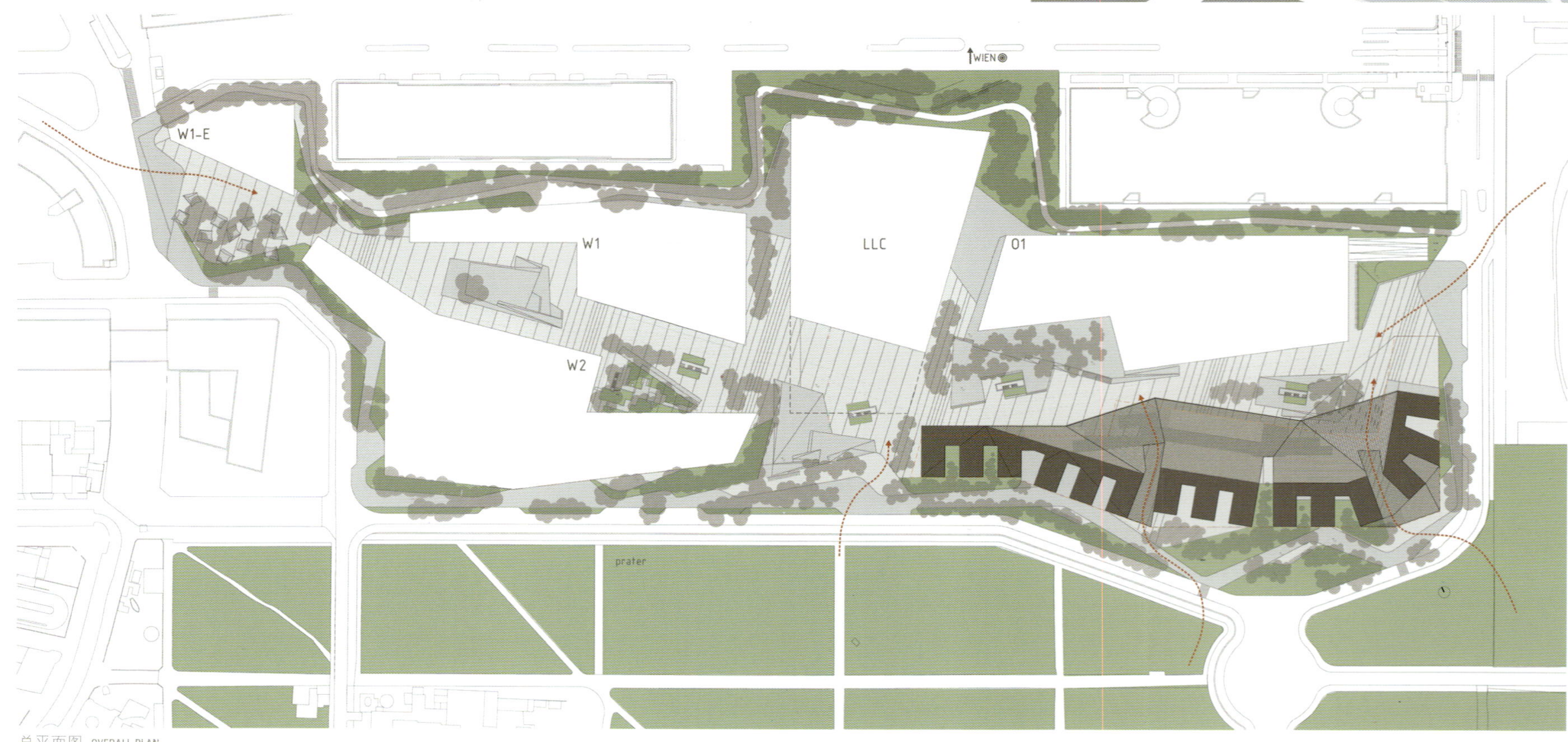

总平面图 OVERALL PLAN

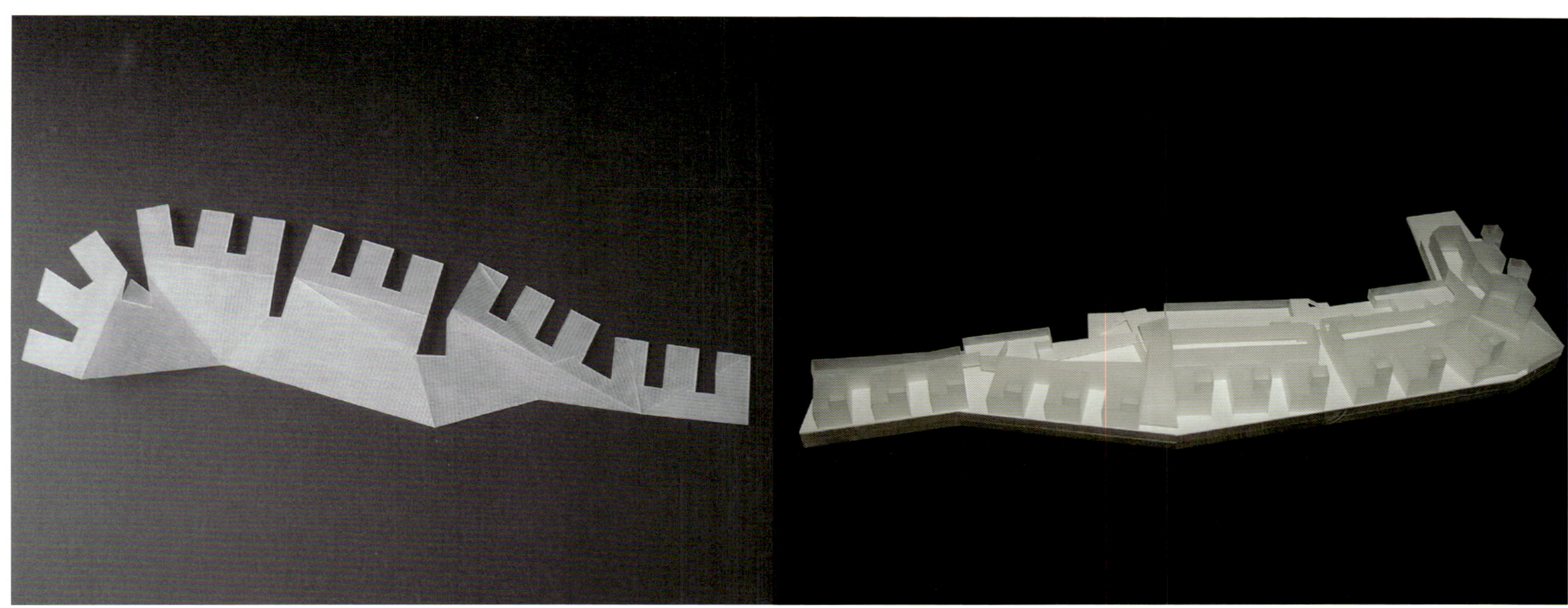

四层平面图 FLOOR PLAN 3

三层平面图 FLOOR PLAN 2

二层平面图 FLOOR PLAN 1

底层平面图 GROUND FLOOR PLAN

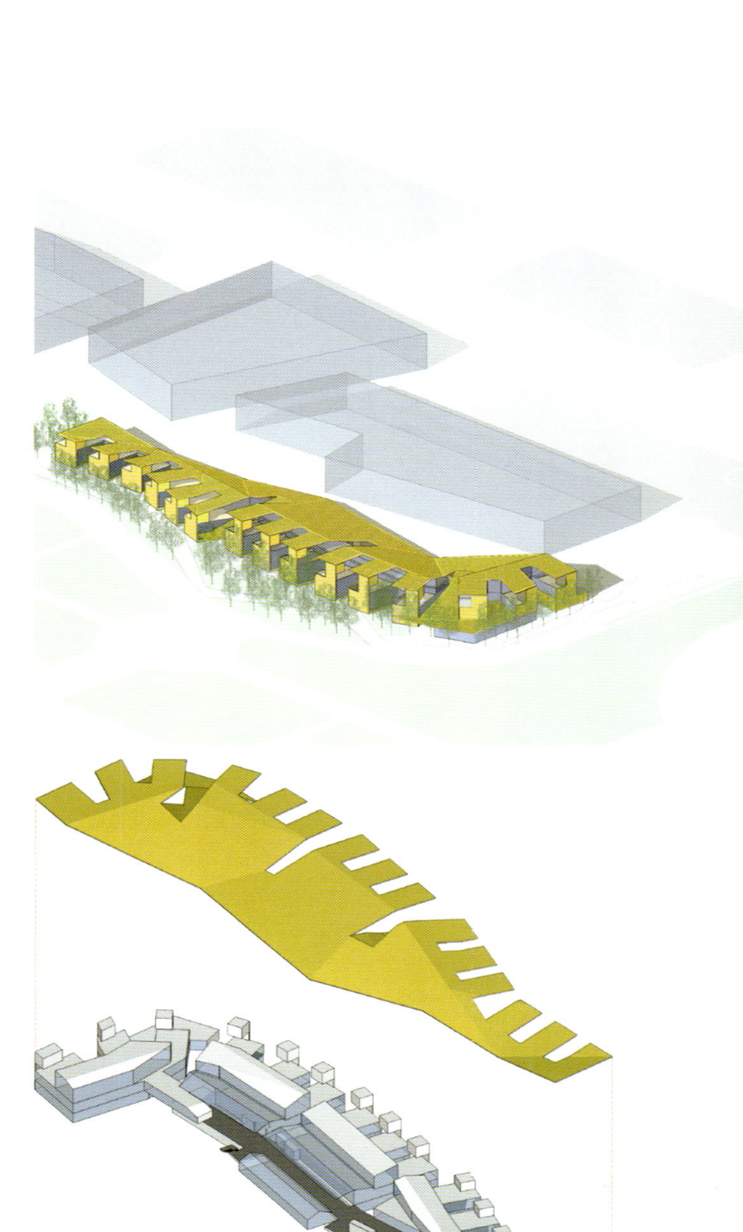

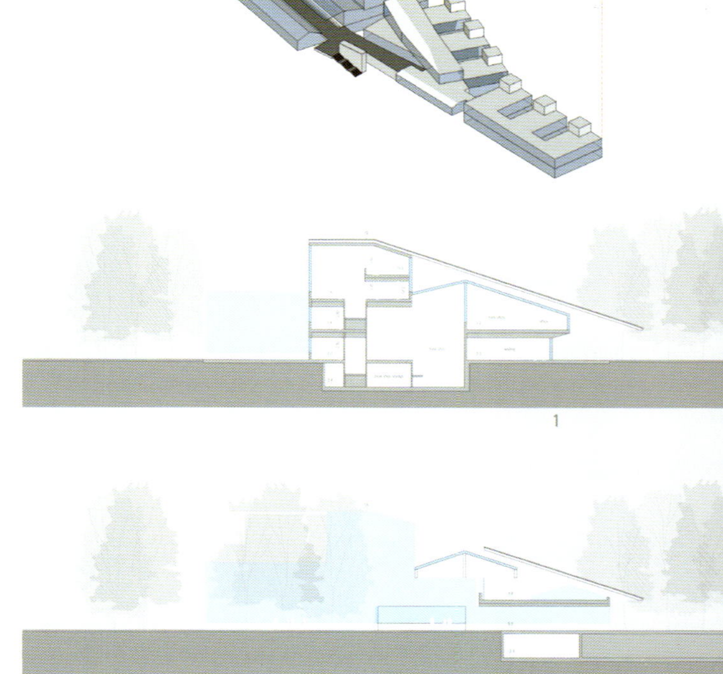

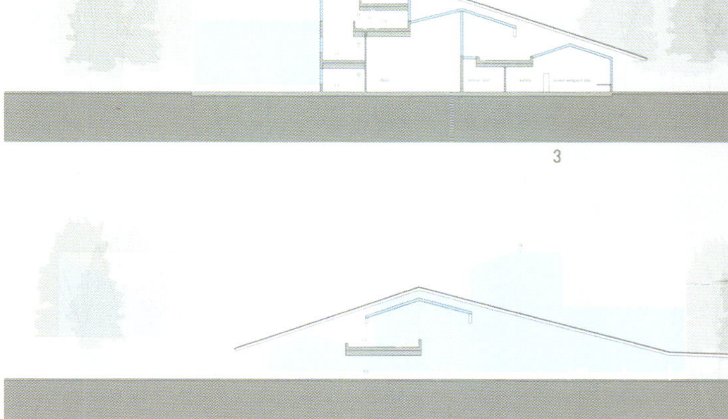

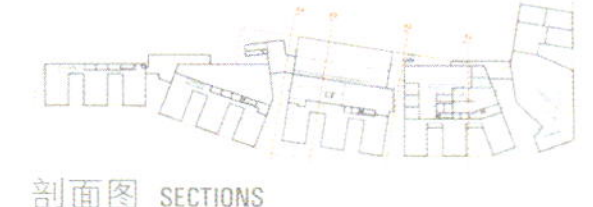

剖面图 SECTIONS

图书馆及学习中心 (LLC)

Library & Learning Center (LLC)

一等奖 · First Prize

Zaha Hadid Architects (建筑师事务所)

Zaha Hadid and Patrik Schumacher (建筑师)

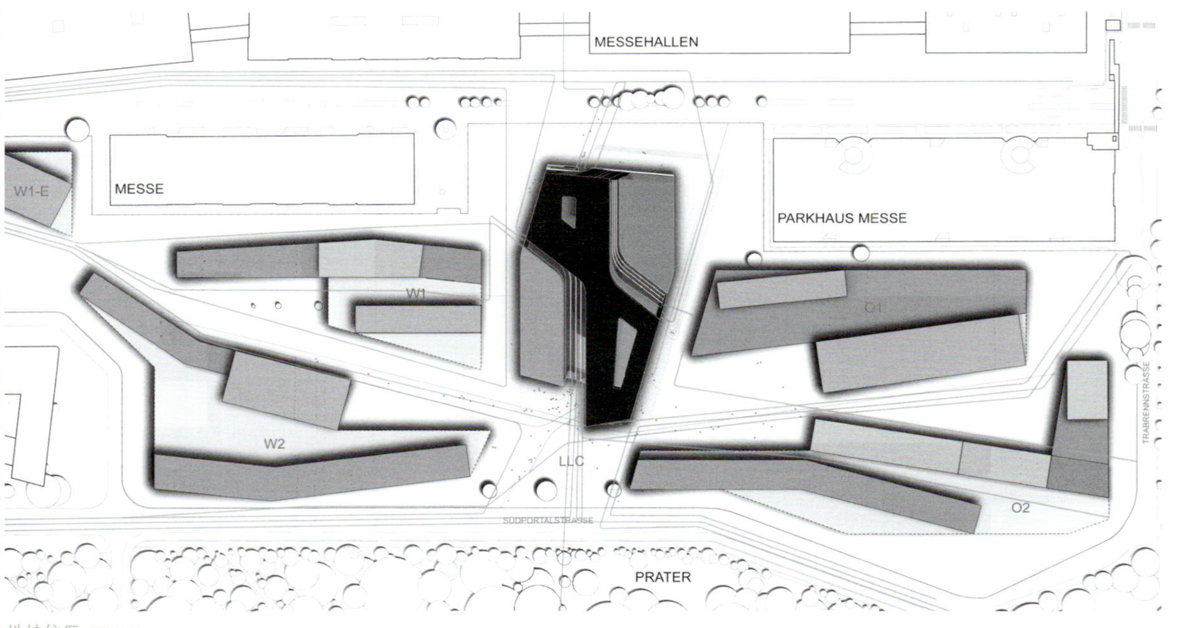

地块位置 SITE PLAN

多边形建筑体

LLC的设计形如一个既有直边又有斜边的立方体。建筑外部的直线向内部延伸时分开，变成流畅的曲线，形成了一个自由成形的内部峡谷，即为中央公共广场。LLC的所有其他设施均位于一个空间中，这个空间也分成了两个独立的带状空间，彼此间互相缠绕，将这个集会空间围在其中。

A POLYGONAL BLOCK

The LLC's design takes the form of a cube with both inclined and straight edges. The straight lines of the building's exterior separate as they move inward, becoming curvilinear and fluid, generating a free-formed interior canyon that serves as the central public plaza. All the other facilities of the LLC are housed within a single volume that also divides, becoming two separate ribbons that wind around each other to enclose this glazed gathering space.

北立面图及西立面图 NORTH AND WEST ELEVATIONS

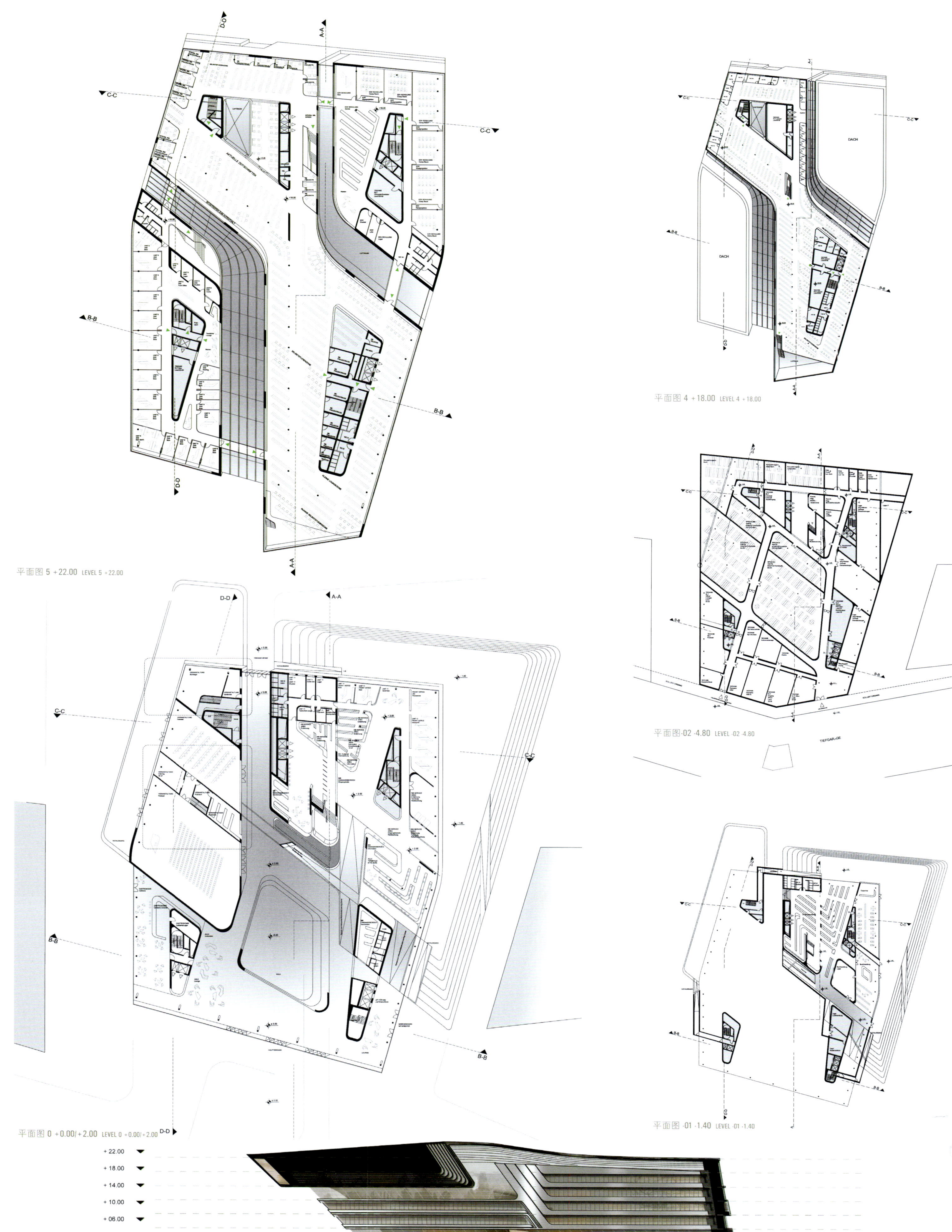

平面图 5 +22.00 LEVEL 5 +22.00

平面图 4 +18.00 LEVEL 4 +18.00

平面图-02 -4.80 LEVEL -02 -4.80

平面图 0 +0.00/+2.00 LEVEL 0 +0.00/+2.00

平面图 -01 -1.40 LEVEL -01 -1.40

东立面图 EAST ELEVATION

图书馆及学习中心 (LLC)

Library & Learning Center (LLC)

二等奖 · Second Prize

Morphosis (建筑师事务所)

Thom Mayne (建筑师)

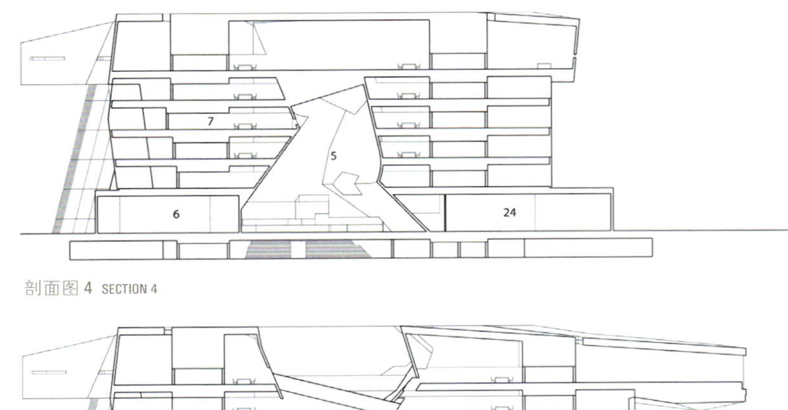

剖面图 4 SECTION 4

剖面图 2 SECTION 2

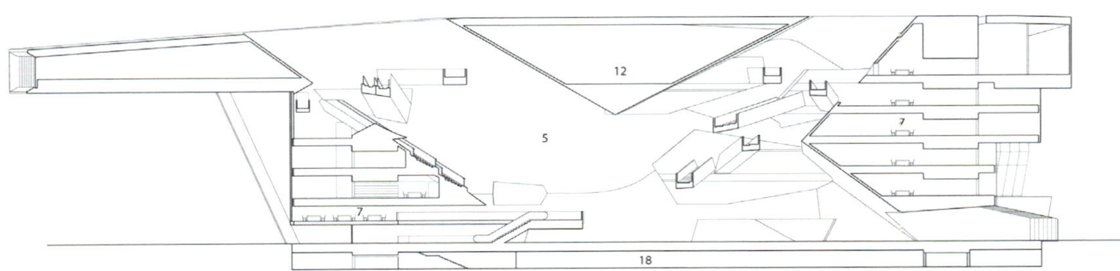

剖面图 1 SECTION 1

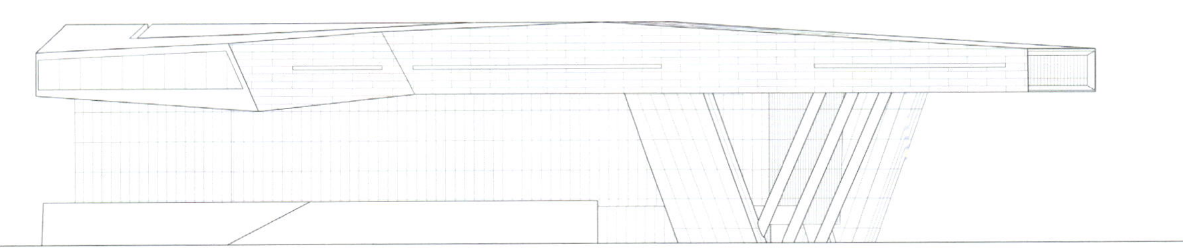

西立面图 WEST ELEVATION

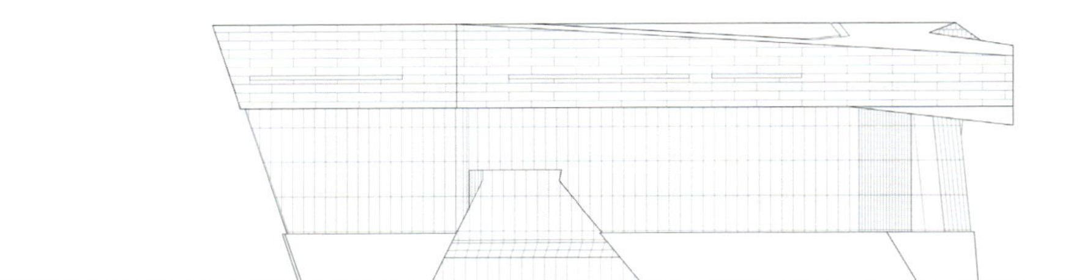

北立面图 NORTH ELEVATION

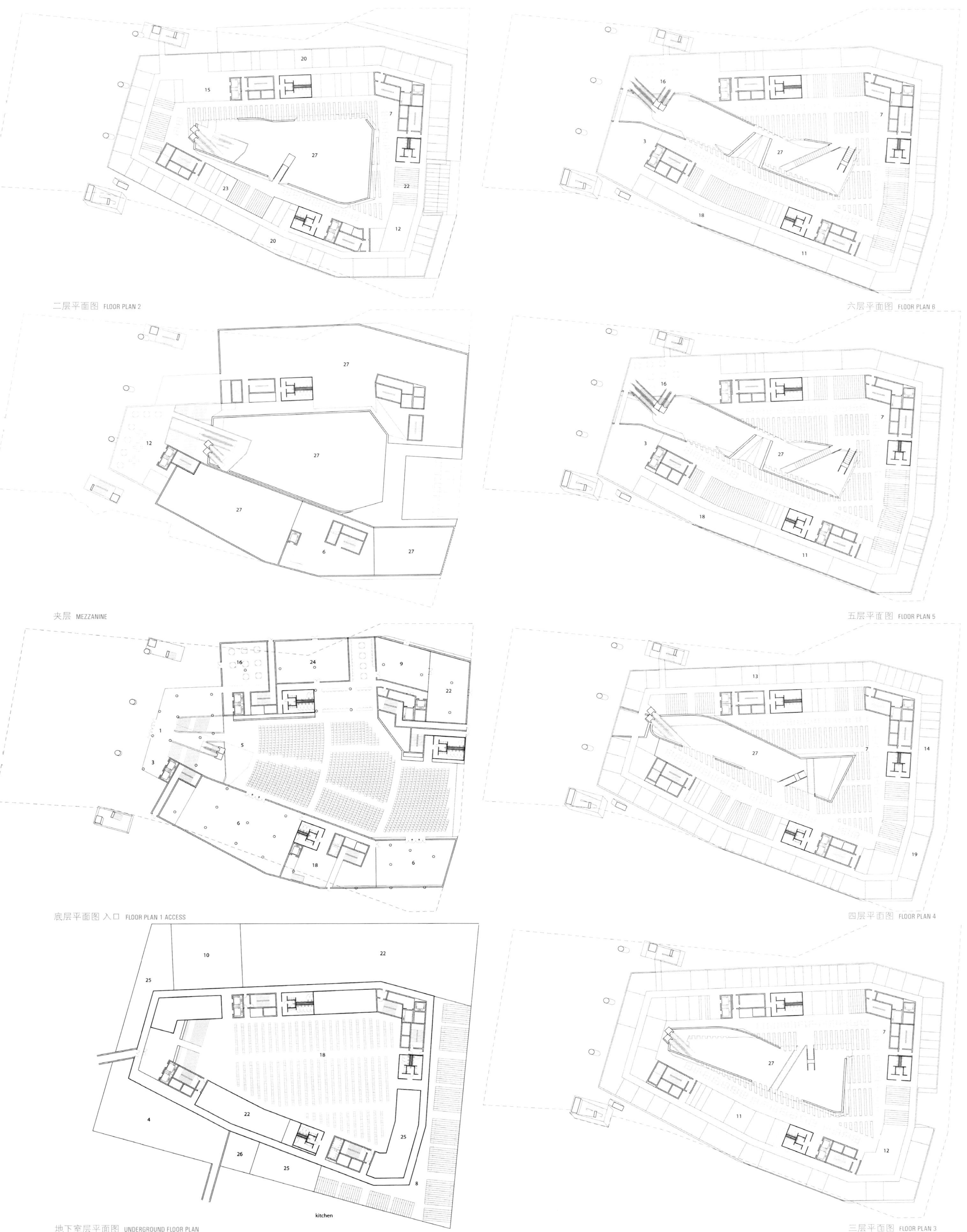

二层平面图 FLOOR PLAN 2

六层平面图 FLOOR PLAN 6

夹层 MEZZANINE

五层平面图 FLOOR PLAN 5

底层平面图 入口 FLOOR PLAN 1 ACCESS

四层平面图 FLOOR PLAN 4

地下室层平面图 UNDERGROUND FLOOR PLAN

三层平面图 FLOOR PLAN 3

部门大楼 (W1)

Department Building (W1)

Carme Pinós (建筑师事务所)

一等奖 · First Prize

总平面图 OVERALL PLAN

分解

运用平行六面体的格局轻松解决了有关办公室位置和项目其他组成部分的问题。面向主干道住宅且人流量最多的立面被改造成了自助餐厅和图书馆，且均设有独立入口。其他立面由30厘米长的垂直遮光栅格做成，作百叶窗之用，同时也方便经常变换图案。

BROKEN

A game of parallelepipeds easily solves the location of offices and the rest of the program. The façade facing the main road houses the most public functions as the cafeteria and library, with an independent access. The façades are made up of 30cm-vertical louvers which function as brisoleil and at the same time introduce a constant change of image.

立面图 ELEVATION

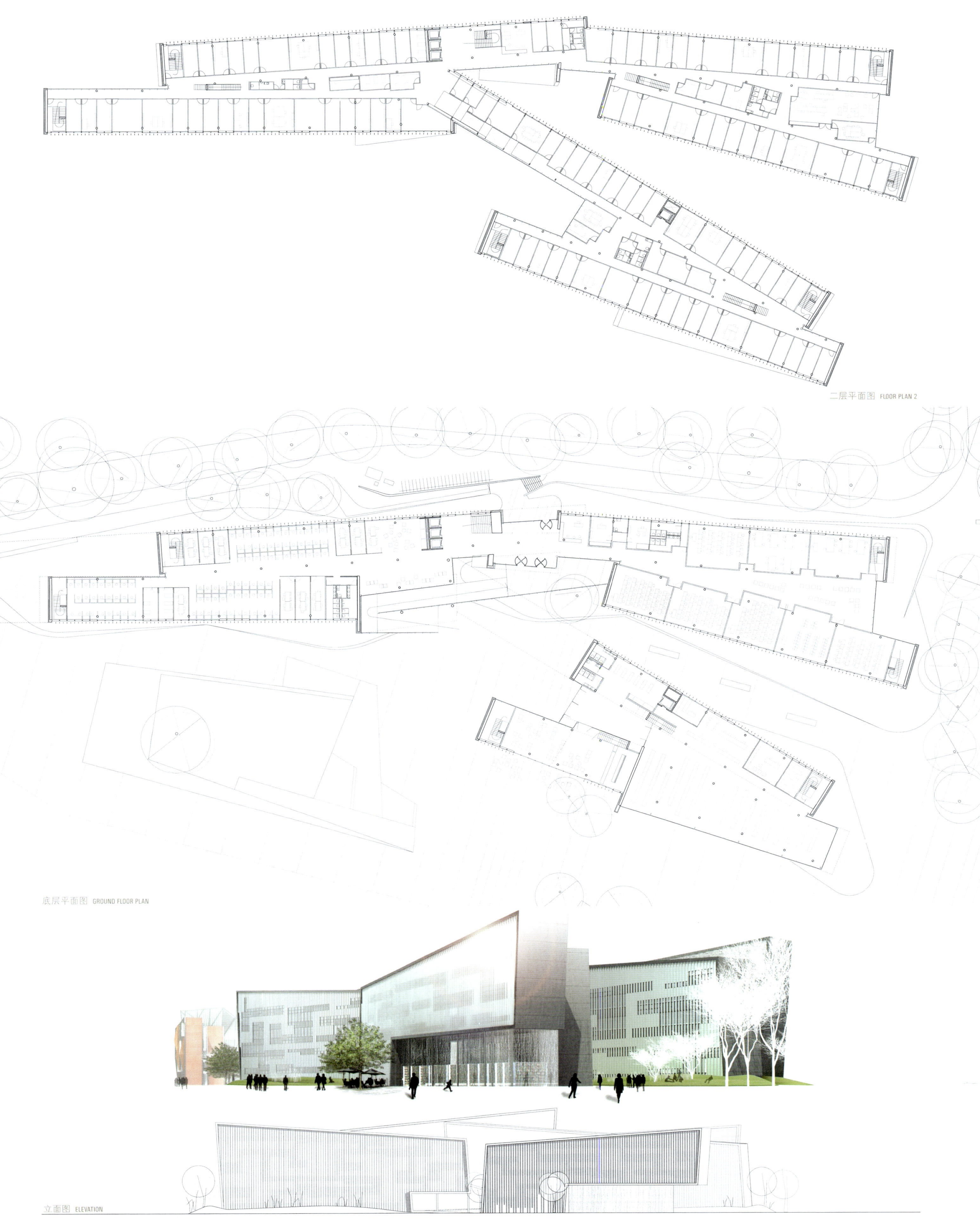

二层平面图 FLOOR PLAN 2

底层平面图 GROUND FLOOR PLAN

立面图 ELEVATION

部门大楼 (W2)

Department Building (W2)

一等奖 · First Prize

CRABstudio (建筑师事务所)

源自西伯利亚落叶松

该建筑的外观柔和、热情、奔放，可以任意地迂回婉转和变换方向。由下往上，颜色从最低点的暗红色渐变为橙色、米色，最终在最高点时变为白色。石板的过滤层取材于西伯利亚落叶松（一种随时间流逝而变白且极为耐用的木材），能遮蔽阳光，并能使建筑本身与普拉特森林的木屏彼此呼应。

FROM SIBERIAN LARCH

The wrapping of the building is seen as a lyrical envelope that can weave and change direction effortlessly. As these climb upwards they will change colour from dark red at the ground level, through orange and cream to white at the highest level. A filter layer of slats – cut from Siberian Larch (a timber that bleaches over time and has a tremendous durability) will enable shading from the sun and architecturally establish a link to the timber screening of the Prater woods.

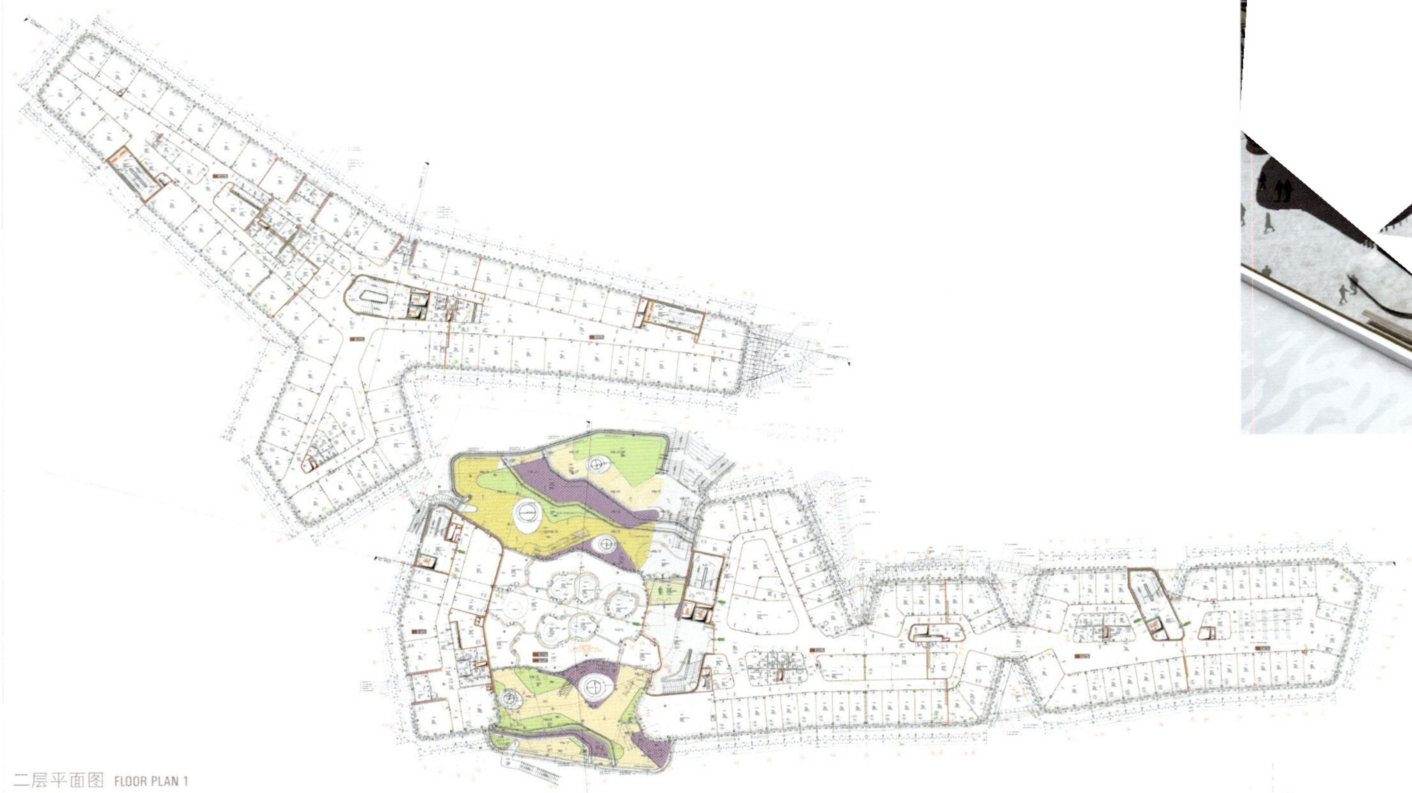

二层平面图 FLOOR PLAN 1

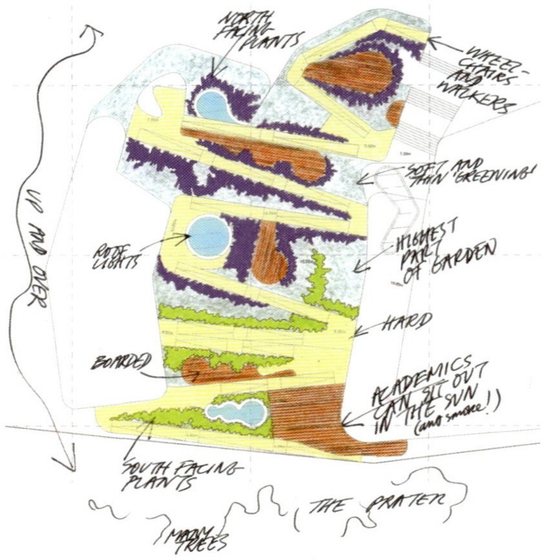

底层平面图 GROUND FLOOR PLAN

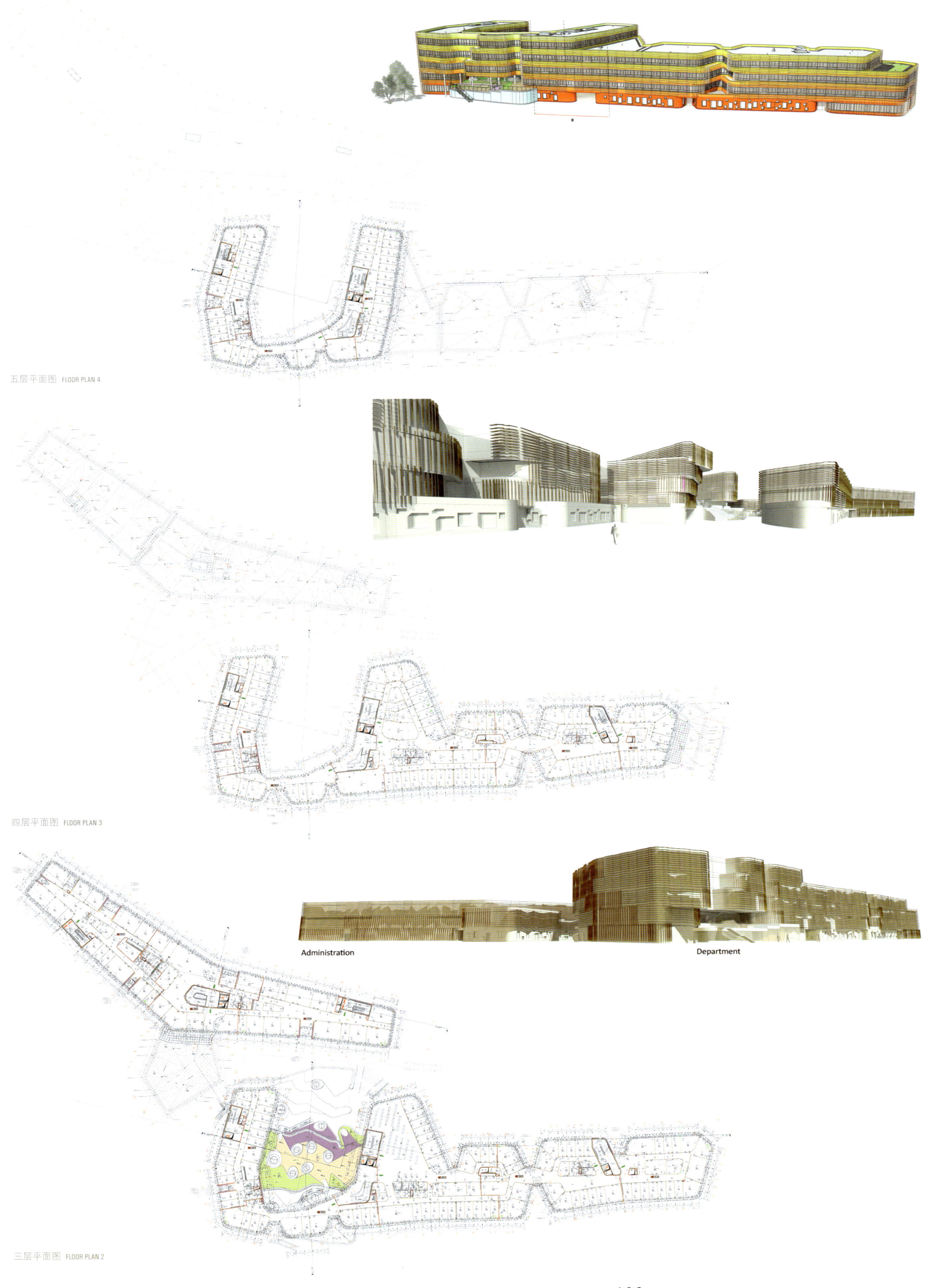

五层平面图 FLOOR PLAN 4

四层平面图 FLOOR PLAN 3

三层平面图 FLOOR PLAN 2

部门大楼 (W2)

Department Building (W2)

二等奖 · Second Prize

Guillermo Vázquez Consuegra (建筑师事务所)

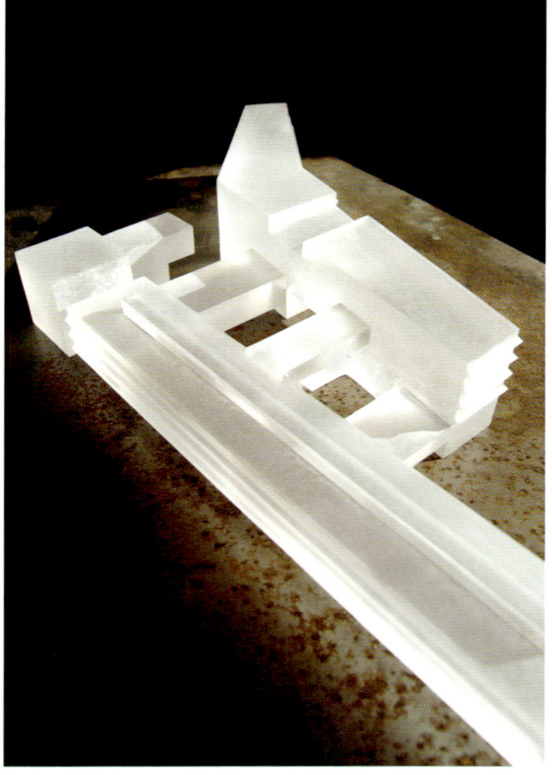

两个建筑体

该工程展现的是两大建筑体。这两个建筑体由一个共有空间分隔开，并呈现一些单个的元素，如小桥、平台、悬桁等，占了一部分空间。

INTERMEDIATE SPACES

The project proposes the construction of two big blocks. Two blocks separated by a common space, characterized by the presence of singular elements: bridges, platforms, cantilevers... which partially cover the space between blocks.

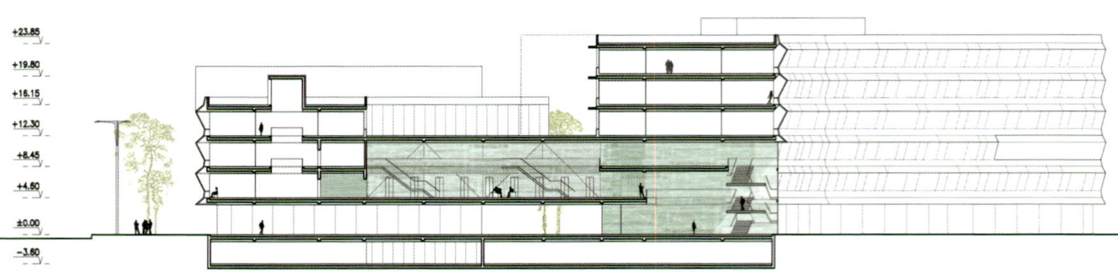

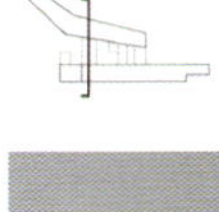

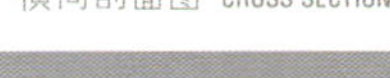

横向剖面图 CROSS SECTION

平面图 +5.00 FLOOR PLAN +5.00

平面图 +0.00 FLOOR PLAN +0.00

纵向剖面图 LONGITUDINAL SECTIONS

部门大楼 (W2)

Department Building (W2)

三等奖 · Third Prize

303055 (竞标代码)

Eric Owen Moss Architects (建筑师事务所)

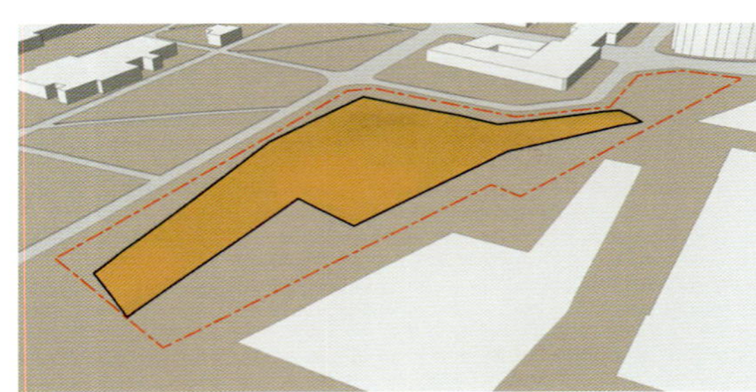

牧场 THE MEADOW

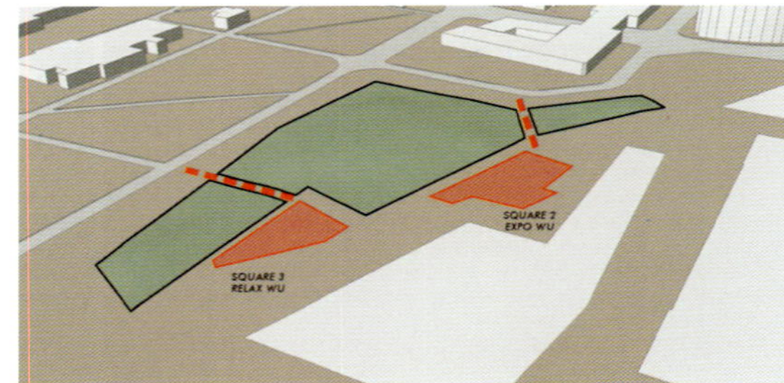

PRATER · 穿行广场 PRATER · SQUARE CROSSWALKS

庭院

在现今的维也纳普拉特和新“漫步公园”之间有一个天然的景观公园。高楼和庭院周围分布着各种行政机构，这些行政机构独立于学术机构之外单独行使职权。天窗和外墙面给行政办公室引入了自然光线。

COURTS

There is a landscaped park between the existing Vienna Prater and the new 'walk along park'. Tower/courts are surrounded by administration uses capable of functioning independently from the academic uses. Skylights and exterior glazing provide natural daylight for the administrative offices.

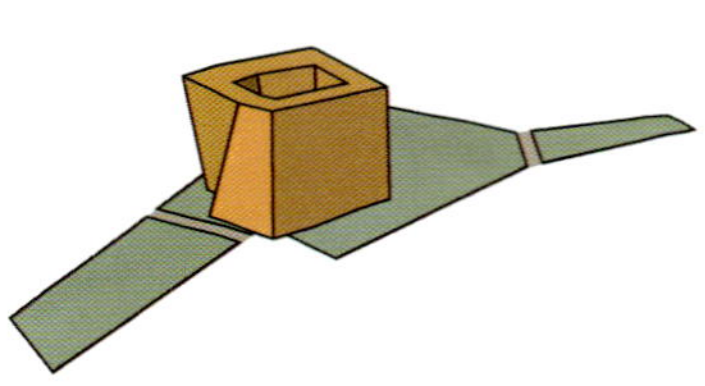

部门 · 夏季庭院
DEPARTMENTS · SUMMER COURT

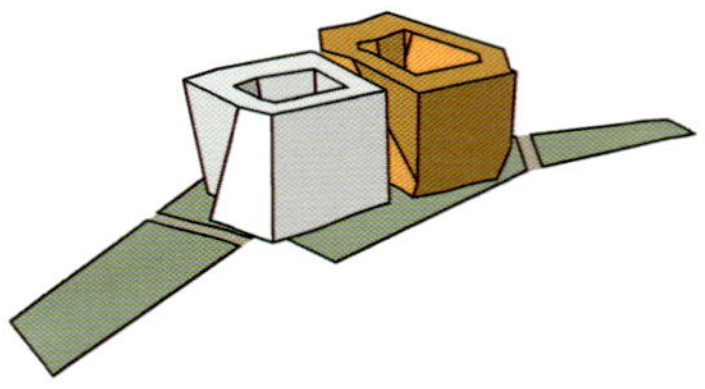

部门 · 冬季庭院
DEPARTMENTS · WINTER COURT

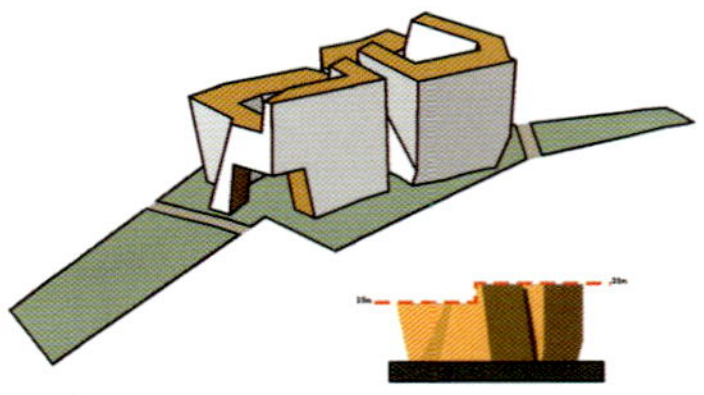

庭院开启及高度限制
COURT OPENINGS AND HEIGHT LIMITS

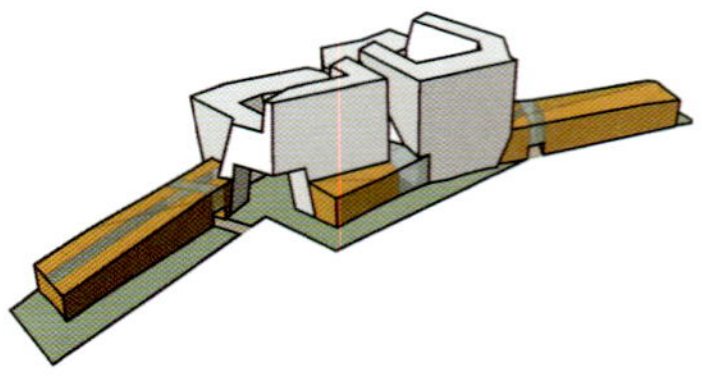

管理部及天台
ADMINISTRATION AND SKYLIGHTS

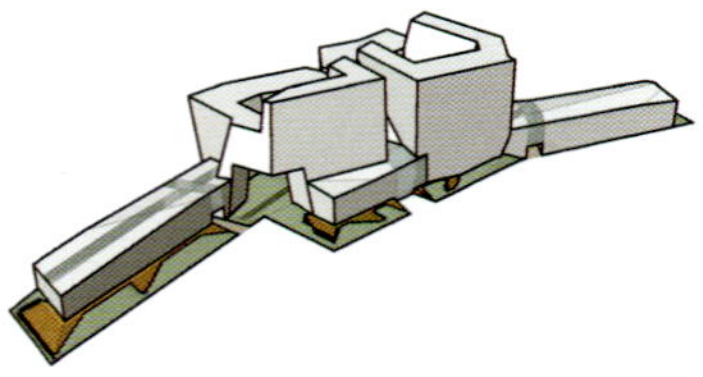

学术广场座位
ACADEMIC SQUARES SEATING

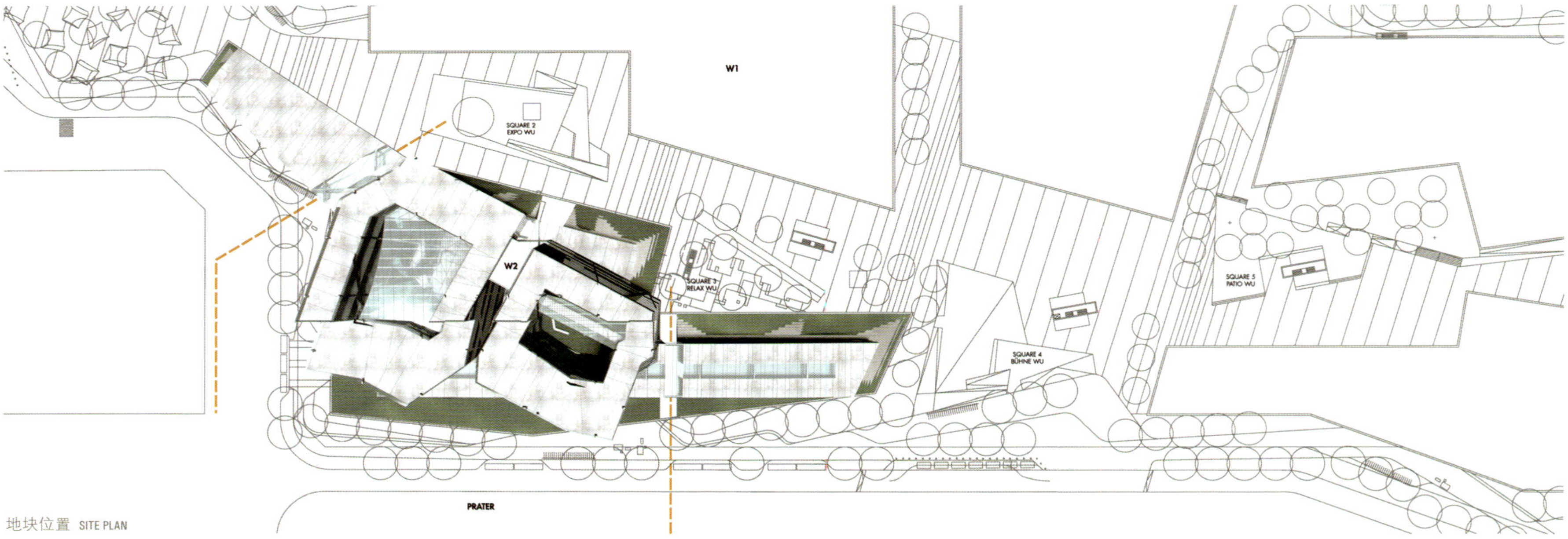

地块位置 SITE PLAN

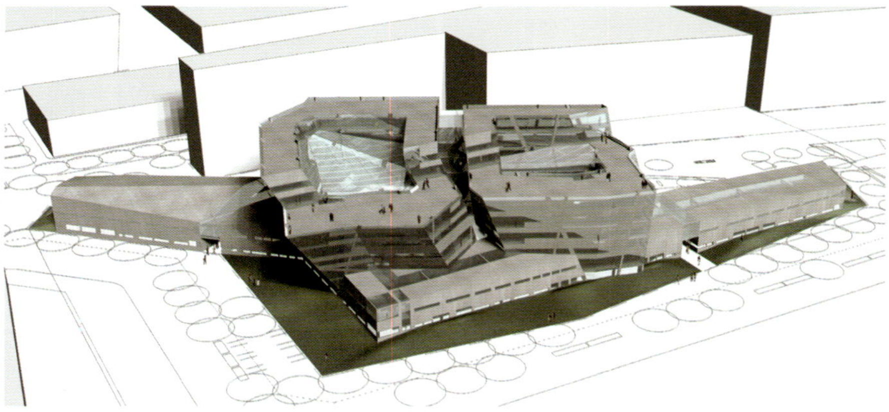

北立面图 NORTH ELEVATION

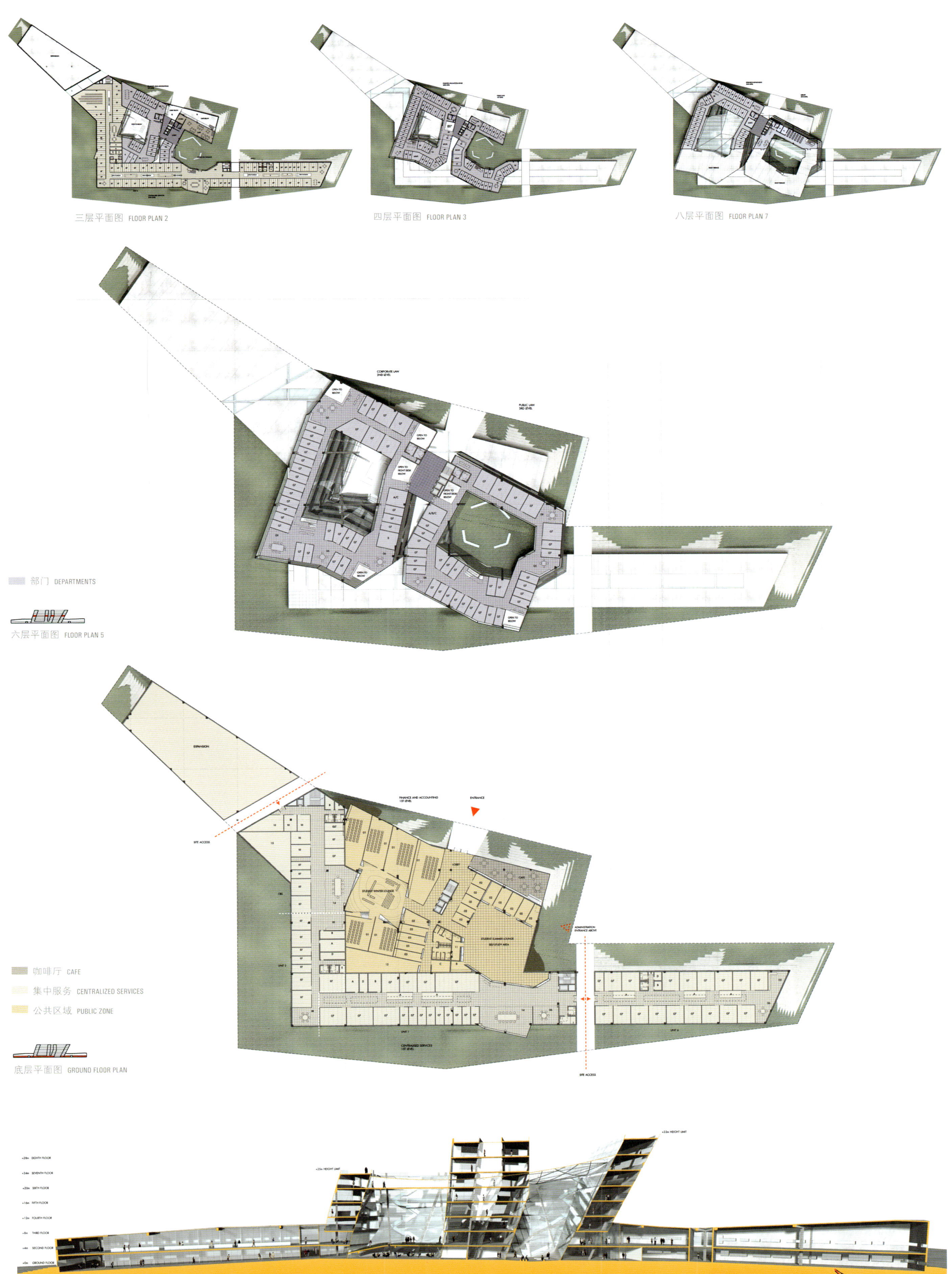
三层平面图 FLOOR PLAN 2
四层平面图 FLOOR PLAN 3
八层平面图 FLOOR PLAN 7
部门 DEPARTMENTS
六层平面图 FLOOR PLAN 5
咖啡厅 CAFE
集中服务 CENTRALIZED SERVICES
公共区域 PUBLIC ZONE
底层平面图 GROUND FLOOR PLAN

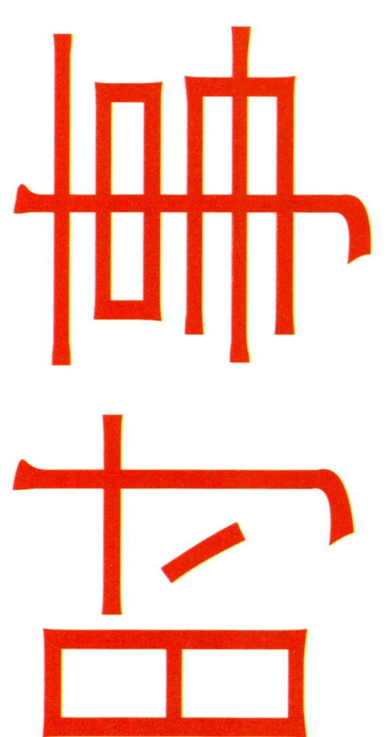

时事 News

文化遗产中心 · 济州岛
Cultural Heritage Center · Korea

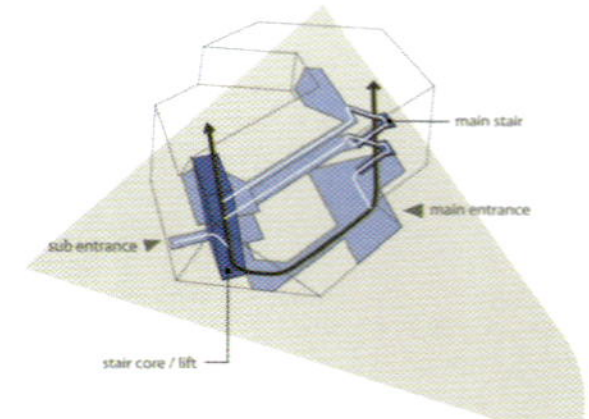

内部流通 INTERNAL CIRCULATION

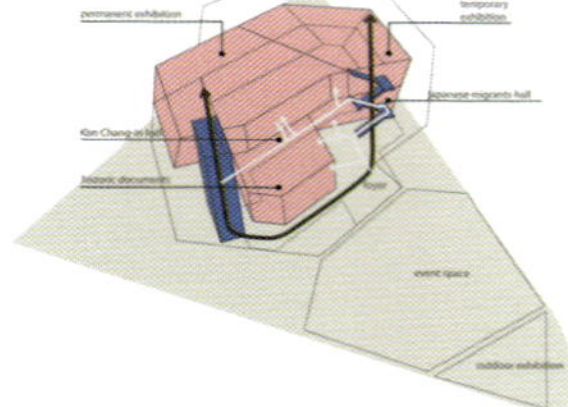

展区空间 EXHIBITION SPACES

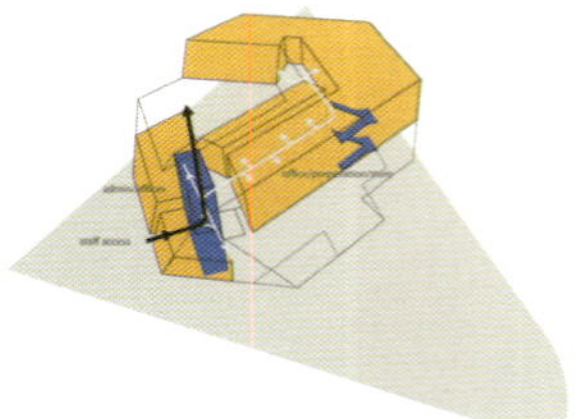

行政区域 ADMINISTRATION SPACES

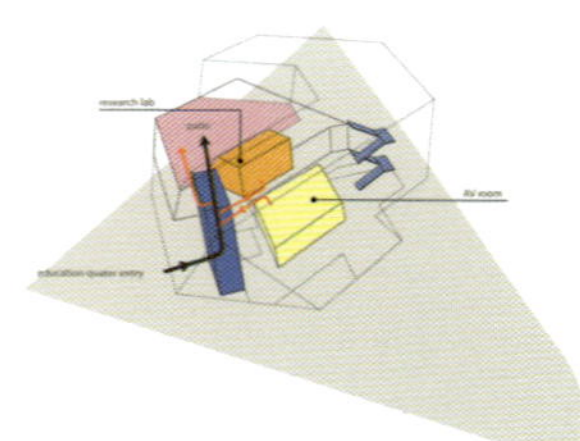

教育区域 EDUCATION SPACES

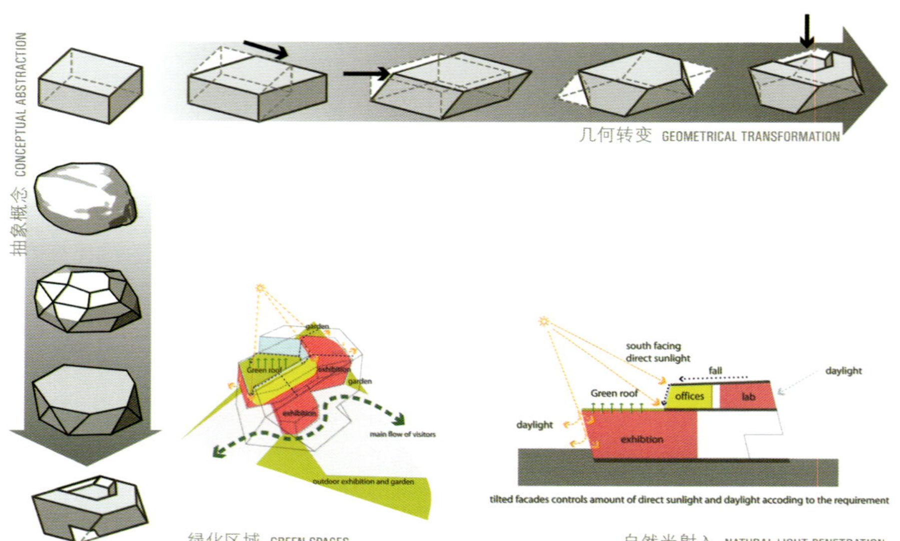

几何转变 GEOMETRICAL TRANSFORMATION

绿化区域 GREEN SPACES

自然光射入 NATURAL LIGHT PENETRATION

由**poly.m.ur+KIM architects**设计的文化遗产中心，坐落于济州岛大学校区的正大门边上。它将成为该大学校区的新标志。受即将在该文化遗产中心陈列展出的文物的启发，设计人员提出了该方案，希望能以之凸显这些文物的意义。

Situated inside of Jeju University campus immediately after the front gate, the Cultural Heritage Centre by **poly.m.ur+KIM architects**, is to become a new face of the University and bring a new identity to the campus. Inspired by the artefacts that are to be exhibited here, the proposal attempts to simulate the significance of these artefacts.

POLY.M.UR + KIM ARCHITECTS (建筑师事务所)
Homin Kim · Chris S. Yoo Hee-kyung Mun · (建筑师)

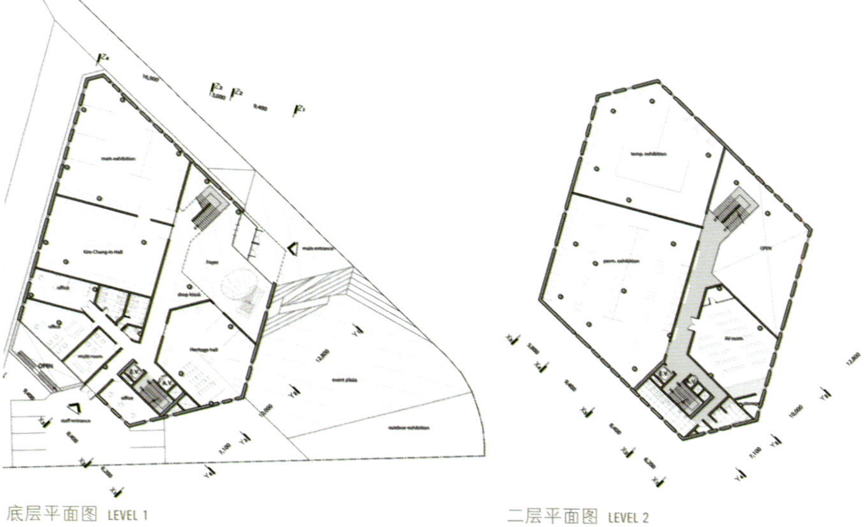

底层平面图 LEVEL 1

二层平面图 LEVEL 2

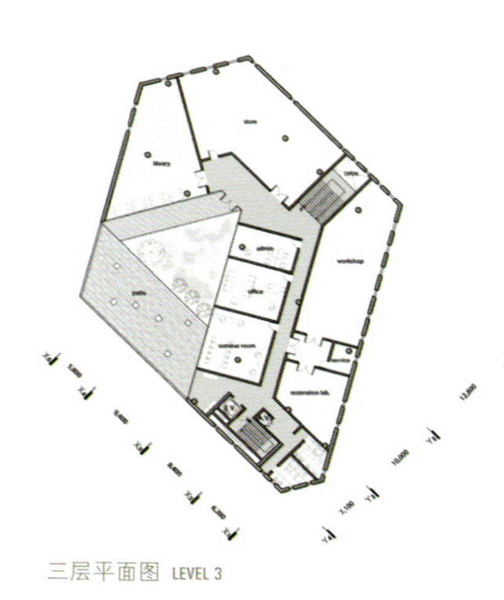

三层平面图 LEVEL 3

剖面图 A SECTION A

剖面图 B SECTION B

会展城·开罗
Expo City · Egypt

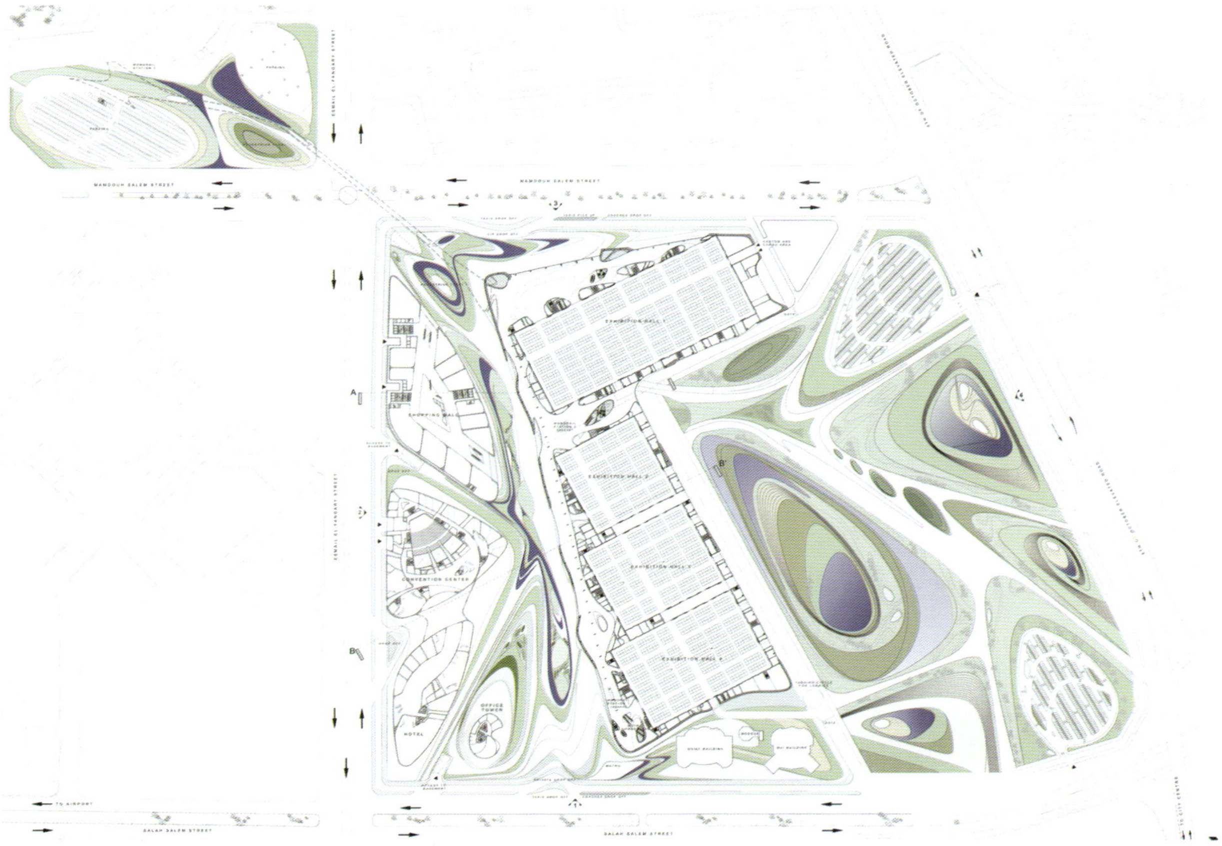

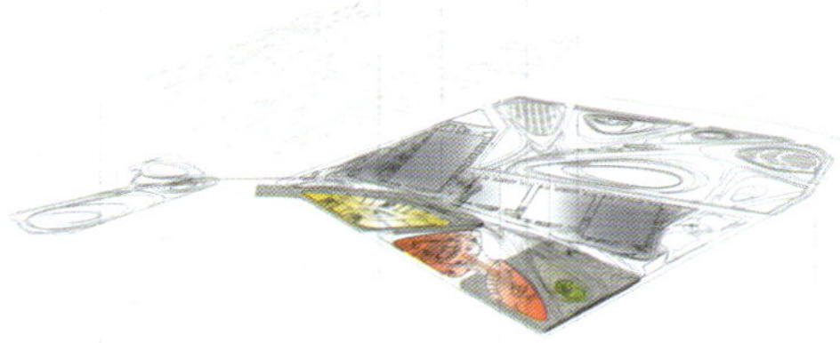

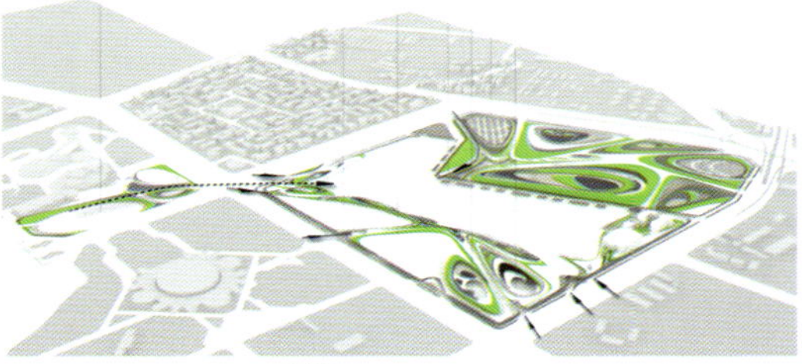

底层平面图 GROUND FLOOR LEVEL

Zaha Hadid Architects从两轮角逐中脱颖而出，将与学识渊博的工程顾问布罗•哈波尔德一起设计开罗会展城。胜出的设计方案展示了一个卓尔不群的城市建筑设施——这个占地45万平方米的会展城位于开罗市中心和机场之间，集会展和会议两大功能为一体，彰显出非凡的艺术气质。

Zaha Hadid Architects has been selected to design the Cairo Expo City, together with multidisciplinary engineering consultancy Buro Happold after a two-phase competition. The winning design delivers a unique facility for the city - a 450,000 square metre, state of the art city for exhibitions and conferences, located between the centre of Cairo and the city's airport.

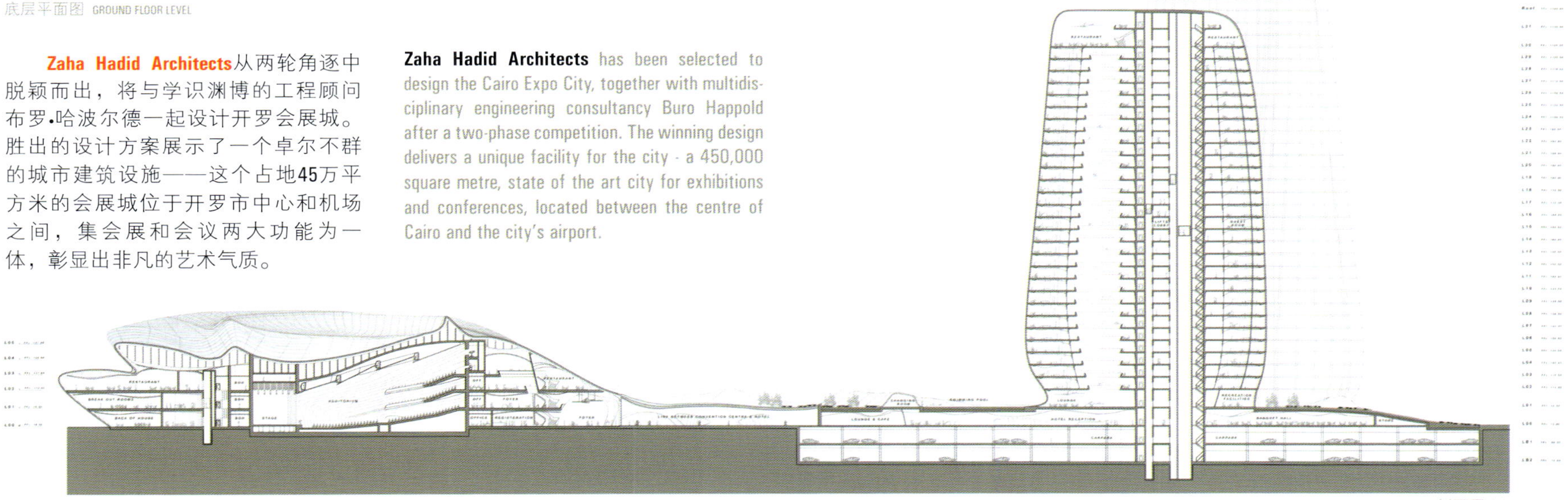

剖面图 AA SECTION AA

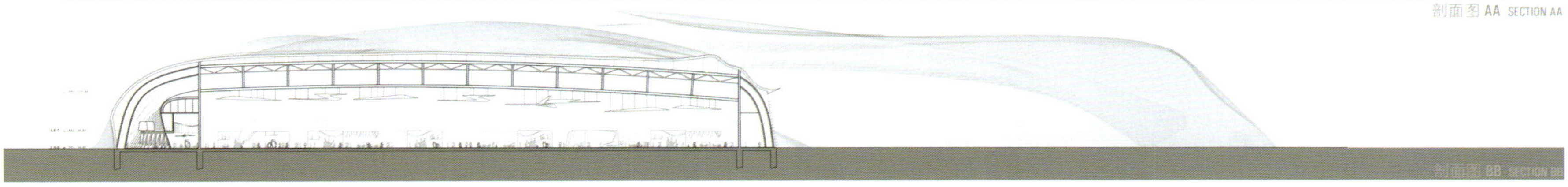

FOYER

EXHIBITION HALL 4

剖面图 BB SECTION BB

新面海都市中心及图书馆 · 奥胡斯
New Central Urban Waterfront Space and Library (MEDIASCAPE) · Denmark

随着对内港沿岸长约500米的地带进行重新规划以及对城市新地标Mediaspace的破土兴建，丹麦第二大城市奥胡斯市的港口处将焕然一新。Schmidt Hammer Lassen architects (SHL Architects)在一场国际竞标中崭露头角，成为"Urban Mediaspace"——斯堪的纳维亚半岛上最大的公共图书馆的设计方。这个投资2.28亿欧元的项目将成为该城市中一个全新的视觉文化焦点。该项目的设计灵感源自港口前端地区的工业化建筑元素，包括烟囱、起重机和仓库。

The port of the second city of Denmark, Arhus, will be revived by the redevelopment of a 500-meter long strip along the inner harbour and the construction of Mediaspace, a new icon of the city. **Schmidt Hammer Lassen architects (SHL Architects)** has won an international competition to design "Urban Mediaspace", the largest public library in Scandinavia. The €228 million scheme will become a new visual and cultural focal point for the city. The project is inspired by the industrial architecture of the harbour front with its chimneys, cranes and warehouses.

一等奖 · FIRST PRIZE
SCHMIDT HAMMER LASSEN ARCHITECTS (建筑师事务所)

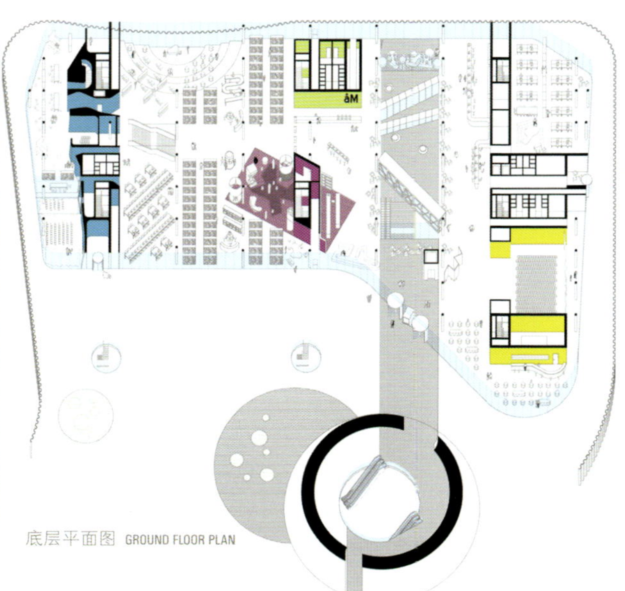

底层平面图 GROUND FLOOR PLAN

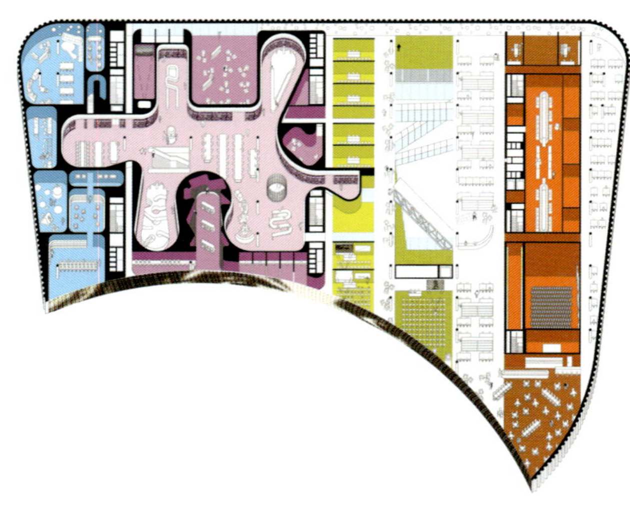

二层平面图 FLOOR PLAN 1

二等奖 · SECOND PRIZE
MECANOO (建筑师事务所)

河西新城区面河开发项目·南京

Nanjing Hexi New Town Riverfront Development · China

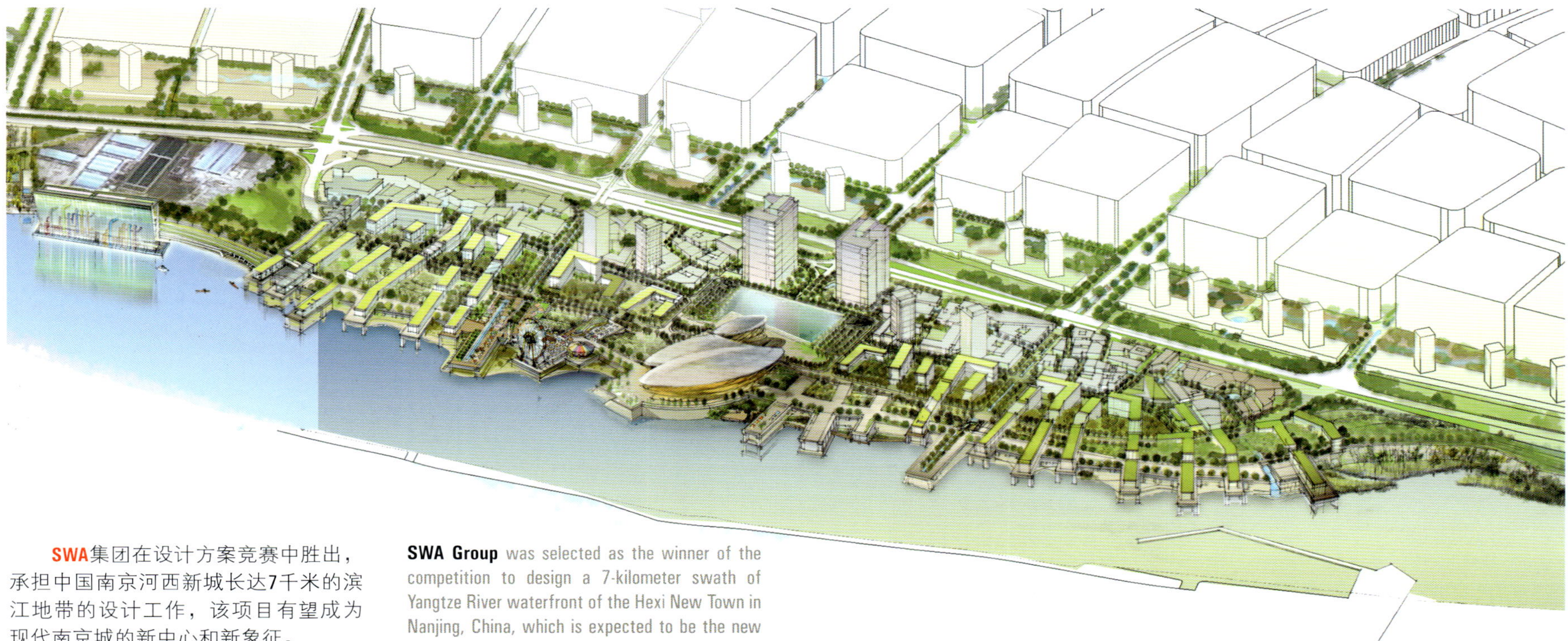

SWA集团在设计方案竞赛中胜出，承担中国南京河西新城长达7千米的滨江地带的设计工作，该项目有望成为现代南京城的新中心和新象征。

SWA Group was selected as the winner of the competition to design a 7-kilometer swath of Yangtze River waterfront of the Hexi New Town in Nanjing, China, which is expected to be the new center and symbol of the modern city of Nanjing.

春季 SPRING

夏季 SUMMER

冬季 WINTER

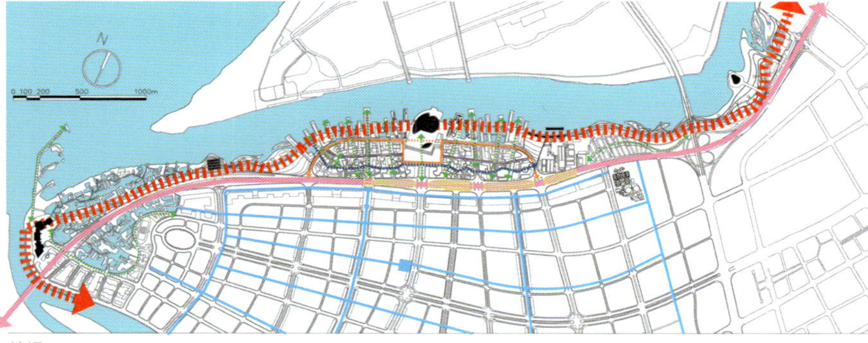

流通 CIRCULATION

开放区域 OPEN SPACES

水文学 HYDROLOGY

概念 CONCEPT

珠海学院新校区 · 中国香港

New campus for Chu Hai College · China

Office for Metropolitan Architecture OMA与**Leigh & Orange Architects**携手在中国香港珠海学院新校区的竞标中一举夺魁。该项目的建筑总面积为2.8万平方米，包含三个学院的教育设施——文学院、理工学院及商学院——共有十个系部和两个研究中心。

The **Office for Metropolitan Architecture OMA** in collaboration with **Leigh & Orange Architects** has won the competition for the new campus for Zhu Hai College of Higher Education in Hong Kong, China. The project, with a gross floor area of 28,000 sqm, consists of education facilities for three faculties – arts, science and engineering, and business – containing 10 departments and two research centres.

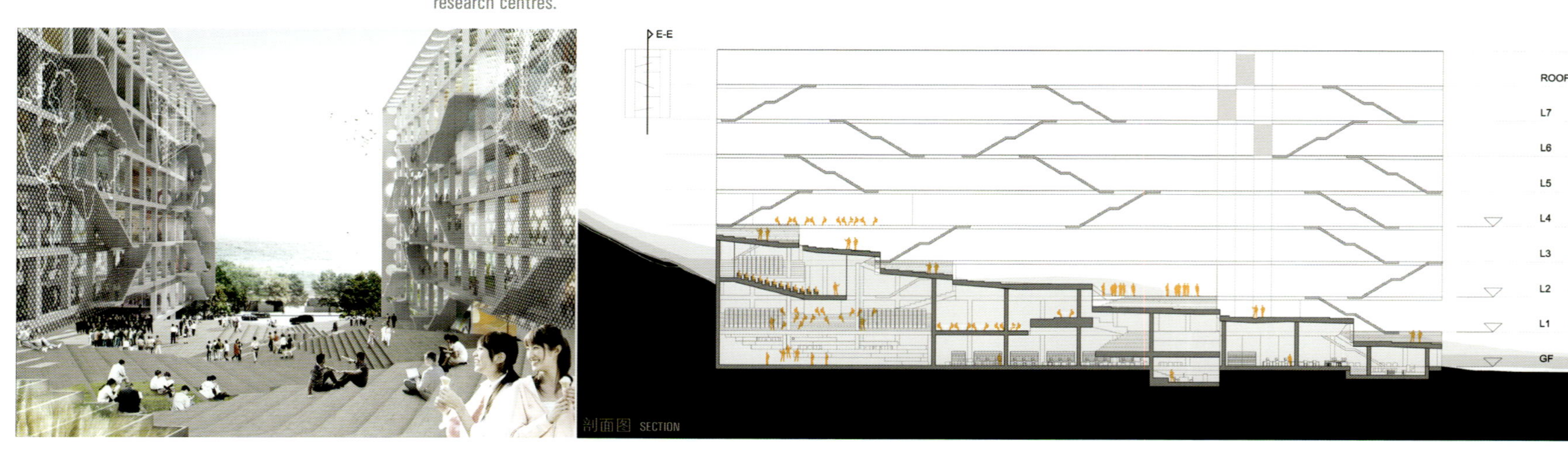

剖面图 SECTION

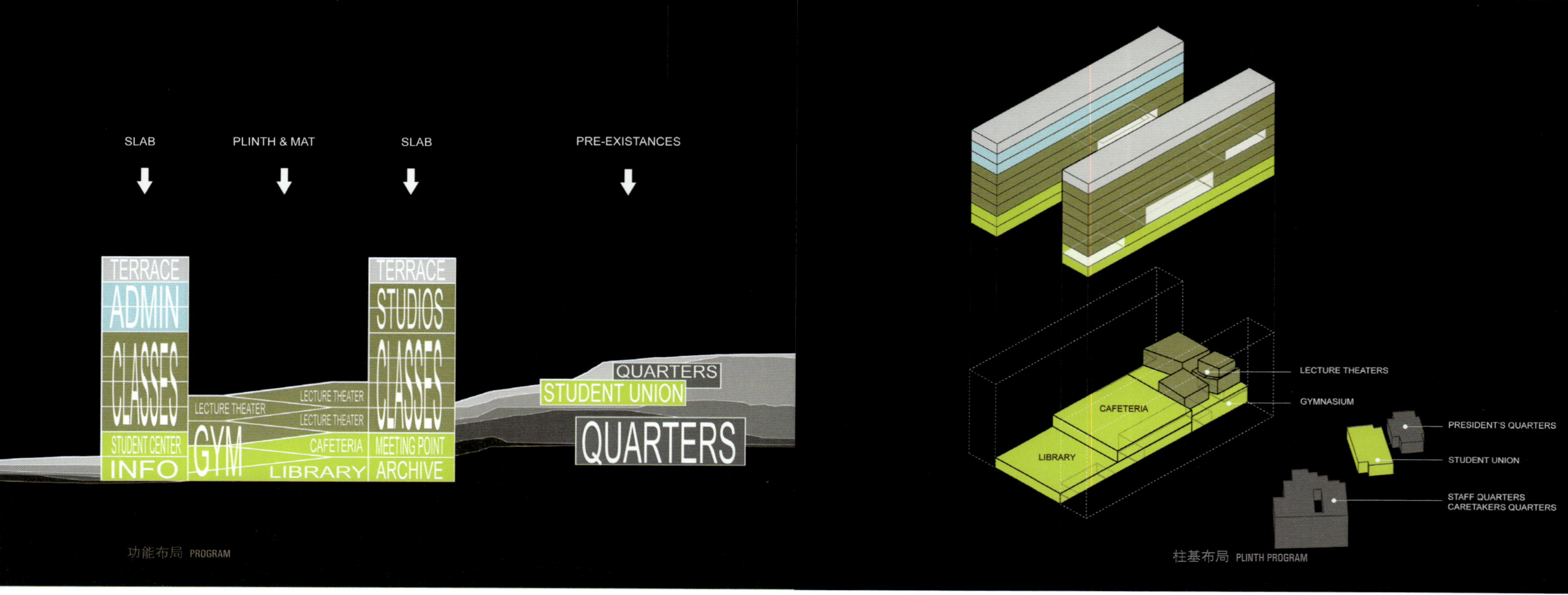

功能布局 PROGRAM

柱基布局 PLINTH PROGRAM

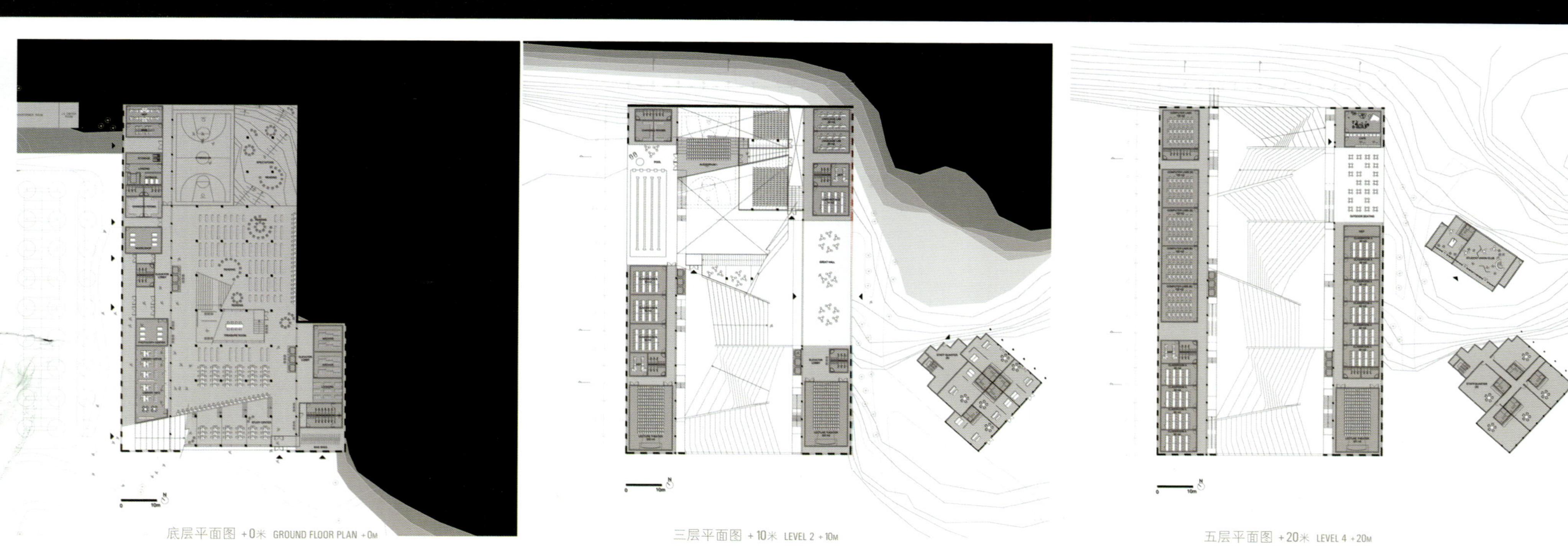

底层平面图 +0米 GROUND FLOOR PLAN +0M

三层平面图 +10米 LEVEL 2 +10M

五层平面图 +20米 LEVEL 4 +20M

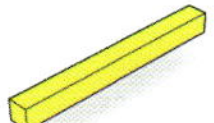 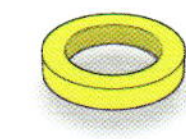 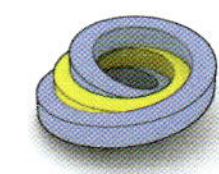 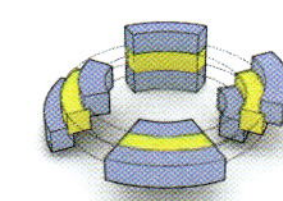 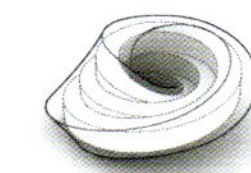

新国家图书馆·阿斯塔纳
New National Library · Kazakhstan

地块位置 SITE PLAN

BIG建筑设计公司在哈萨克斯坦首都阿斯塔纳新国家图书馆的国际公开设计竞赛中荣获一等奖。这幢新建筑占地3.3万平方米，分布在一个连续循环的莫比乌斯环上，由两组环环相扣的完美圆形和公共盘旋空间构成。

BIG Bjarke Ingels Group, has been awarded with the first prize on an open international design competition for Kazakhstan's new National Library in Astana. The new building has an area of 33.000 sqm, arranged as a continuous circulation on a Möbius Strip, as the result of 2 interlocking structures: the perfect circle and the public spiral.

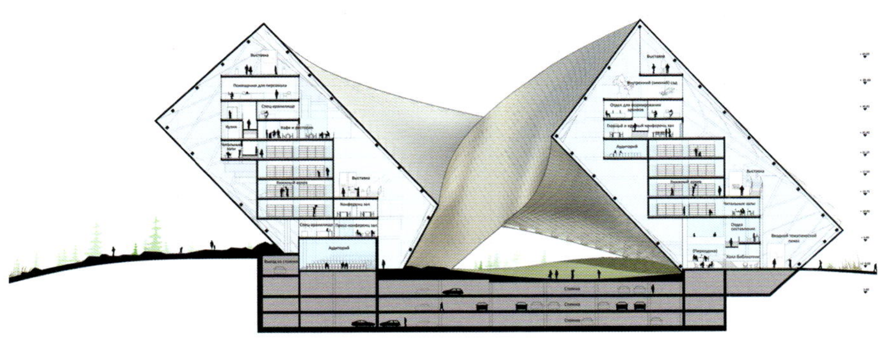

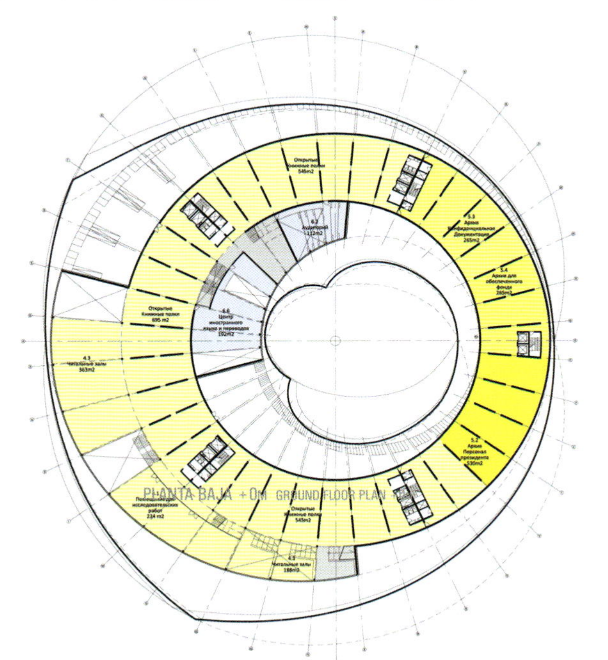

七层平面图 FLOOR PLAN 6

公寓开发·斯德哥尔摩
Residential development · Sweden

3XN建筑事务所在设计斯德哥尔摩的住宅开发项目——Vällingby Parkstad的国际邀标竞赛中拔得头筹。该建筑呈现曲线型设计，生动地塑造了各类造型的阳台。整幢建筑视野开阔，如同空中花园一般。该项目计划于2011年动工。

3XN has won the first prize in an invited international competition to design the new structure marking the entrance to Vällingby Parkstad - a residential development in Stockholm. The building's curved design embraces the area and the lively shaped balconies opens up the structure towards the surroundings thus raising the park up in the air. Construction is planned to begin in 2011.

图书预购

pre-order

7 本 7 books

630¥ 含税及邮资 taxes and shipping costs included

570¥ 学生优惠价 special prize for students 含税及邮资 taxes and shipping costs included

我想从第几辑开始购买未来建筑系列图书
I wish to subscribe to *future* arquitecturas starting from series number..........

预购信息 *pre-order information*

姓名 NAME ....

身份证号码 ID NUMBER ....

单位 COMPANY ....

地址 ADDRESS ....

....

邮编 ZIP CODE ....

电话 PHONE ....

电邮 E-MAIL ....

腾讯 QQ ....

更多信息请联系 For more information contact:

未来建筑中国联络办公室 *future* arquitecturas s.l. China
杭州市文晖路303号交通大厦11楼
邮编：310014

china@arqfuture.com
QQ 860464402

日期 DATE 签名 SIGNATURE

www.arqfuture.com

2010年有效 valid only for year 2010

future 未来
ARQUITECTURAS 建筑

主编 · directors and publishers
Gerardo Mingo Pinacho · Gerardo Mingo Martínez (西)
联合主办单位 · co-sponsors
浙江大学建筑工程学院 · ZUCCEA
浙江大学建筑设计研究院 · ADRZU
西班牙未来建筑 · future arquitecturas s.l.
联合出版 · co-publishers
浙江大学出版社 Zhejiang University Press
图形 · layout
Carlos de Navas Paredes (西)
执行编辑 · managing editor
Gabriela Vélez Trueba (西) · gabriela@arqfuture.com
中国地区公司合伙人 · corporate partner in China
赵磊 Zhao Lei · leizhao@arqfuture.com
美洲地区公司合伙人 · corporate partner in America
Santiago Vélez (厄) · svelez@arqfuture.com
行政人员 · administration
Belén Carballedo (西) · belen@arqfuture.com
Ignacio Rodríguez (西) · future@arqfuture.com
销售部 · distribution department
曾江福 Zeng Jiangfu
手机 cell phone: 13564489269
电话 telephone: 20 65877188
广告 · advertising
china@arqfuture.com